$$\mathcal{D}_{ij} = \mathcal{D}_{ji}$$

$$\frac{\partial c_i}{\partial t} + \mathbf{v} \cdot \nabla c_i = \mathcal{D} \nabla^2 c_i$$

The Newman Lectures on Transport Phenomena

The Newman Lectures on Transport Phenomena

John Newman
Vincent Battaglia

Published by

Jenny Stanford Publishing Pte. Ltd.
Level 34, Centennial Tower
3 Temasek Avenue
Singapore 039190

Email: editorial@jennystanford.com
Web: www.jennystanford.com

British Library Cataloguing-in-Publication Data
A catalogue record for this book is available from the British Library.

The Newman Lectures on Transport Phenomena

ISBN 978-981-4774-27-7 (Hardcover)
ISBN 978-1-315-10829-2 (eBook)

Contents

Introduction ix

SECTION A: BASIC TRANSPORT RELATIONS

1. Conservation Laws and Transport Laws **3**

2. Fluid Mechanics **5**
2.1 Conservation of Mass 5
2.2 Conservation of Momentum 6
2.3 Momentum Flux 8
2.4 Assumptions 9

3. Microscopic Interpretation of the Momentum Flux **11**

4. Heat Transfer in a Pure Fluid **13**

5. Concentrations and Velocities in Mixtures **19**

6. Material Balances and Diffusion **23**

7. Relaxation Time for Diffusion **27**

8. Multicomponent Diffusion **31**

9. Heat Transfer in Mixtures **39**

10. Transport Properties **41**

11. Entropy Production **47**

12 Coupled Transport Processes **53**
12.1 Entropy Production 56
12.2 Thermoelectric Effects 58
12.2.1 Energy Transfer 60
12.2.2 Thermoelectric Equation 61
12.2.3 Heat Generation at an Interface 62

12.2.4 Heat Generation in the Bulk 63
12.2.5 Thermoelectric Engine 63
12.2.6 Optimization 65
12.3 Fluctuations and Microscopic Reversibility 67
12.3.1 Macroscopic Part 68
12.3.2 Ensemble Averages 70
12.3.3 Microscopic Reversibility and Probability of States 71
12.3.4 Decay of Fluctuations 73
12.3.5 Summary 74

SECTION B: LAMINAR FLOW SOLUTIONS

13. Introduction **83**

14. Simple Flow Solutions **87**
14.1 Steady Flow in a Pipe or Poiseuille Flow 87
14.2 Couette Flow 88
14.3 Impulsive Motion of a Flat Plate 88

15. Stokes Flow past a Sphere **95**

16. Flow to a Rotating Disk **101**

17. Singular-Perturbation Expansions **107**

18. Creeping Flow past a Sphere **117**

19. Mass Transfer to a Sphere in Stokes Flow **123**

20. Mass Transfer to a Rotating Disk **131**

21. Boundary-Layer Treatment of a Flat Plate **135**

22. Boundary-Layer Equations of Fluid Mechanics **141**

23. Curved Surfaces and Blasius Series **147**

24. The Diffusion Boundary Layer **153**

25. Blasius Series for Mass Transfer **167**

26. Graetz–Nusselt–Lévêque Problem **173**
26.1 Solution by Separation of Variables 174
26.2 Solution for Very Short Distances 176
26.3 Extension of Lévêque Solution 177
26.4 Mass Transfer in Annuli 178

27. Natural Convection **183**

28. High Rates of Mass Transfer **189**

29. Heterogeneous Reaction at a Flat Plate **197**

30. Mass Transfer to the Rear of a Sphere in Stokes Flow **207**

31. Spin Coating **217**

32. Stefan–Maxwell Mass Transport **223**

SECTION C: TRANSPORT IN TURBULENT FLOW

33. Turbulent Flow and Hydrodynamic Stability **251**
33.1 Time Averages of Equations of Motion, Continuity, and Convective Diffusion 251
33.2 Hydrodynamic Stability 252
33.3 Eddy Viscosity, Eddy Diffusivity, and Universal Velocity Profile 252
33.4 Application of These Results to Boundary Layers 252
33.5 Statistical Theories of Turbulence 253

34. Time Averages and Turbulent Transport **255**

35. Universal Velocity Profile and Eddy Viscosity **261**

36. Turbulent Flow in a Pipe **265**

37. Integral Momentum Method for Boundary Layers **269**

38. Use of Universal Eddy Viscosity for Turbulent Boundary Layers **273**

39. Mass Transfer in Turbulent Flow **275**

40. Mass Transfer in Turbulent Pipe Flow **281**

41. Mass Transfer in Turbulent Boundary Layers **287**

42. New Perspective in Turbulence **293**

Appendix A: Vectors and Tensors 297
Appendix B: Similarity Transformations 303

Index 309

Introduction

This book presents fluid mechanics and heat and mass transfer from a fundamental viewpoint. It gives quantitative material which permits senior undergraduate and graduate students to develop models describing system behavior. The presentation is broken into three parts: Basic Transport Relations, Laminar Flow Solutions, and Transport in Turbulent Flow.

The first part lays out the governing physical laws, in some detail. The detail is useful so that one can formulate a mathematical description of a given problem and at the same time realize what approximations are being made. Thus, one can get a good idea of system behavior with a simple model, but if this is not adequate for the purpose, one can refine the model to include other factors that are essential to describe the relevant features of the system for the present application.

The second part seeks to solve the developed model so as to yield quantitative results of system behavior. Some of these are really simple, and the reader is likely to be already familiar with the results. Other examples treat systems that have more complicated behavior but are still tractable. The reader is likely to want to go beyond this and explore a problem that has not been treated before. Then, the first step is to formulate the problem, with differential equations and appropriate boundary conditions, on the basis of the material in Part A. The second step is to solve the mathematical problem, numerically if necessary, using tools developed in Part B. The third part of the problem is to contemplate the meaning of the result. This serves to refine one's intuition in addition to getting a specific result, thereby gaining insight into unfamiliar but related systems.

The third part reviews empirical and semi-empirical treatments of turbulent systems. While Part B treats systems that follow straightforward physical laws, chaotic systems need to use experimental observations to augment the basic physical laws themselves. Turbulent systems are too important to be ignored, and tools should be refined even today. This part includes a brief chapter outlining a new approach.

Transport phenomena is closely related to other fields. It draws heavily on mathematics. *The Newman Lectures on Mathematics* [1] reviews important mathematical methods. Since similarity transformations have been used a lot in this book, a chapter on that subject is borrowed from Ref. [1] and placed in Appendix B. Students may not be familiar with singular perturbations; this mathematical topic is developed as a useful tool in this book.

This book on transport phenomena treats real systems and draws heavily on thermodynamics. Students should find *The Newman Lectures on Thermodynamics* [2] a useful resource in this area. Charged species are involved in some topics treated here in this book. Our book *Electrochemical Systems* [3] provides an appropriate background and is a thorough resource.

References

1. John Newman and Vincent Battaglia. *The Newman Lectures on Mathematics.* Singapore: Jenny Stanford Publishing (formerly Pan Stanford Publishing), 2018.
2. John Newman and Vincent Battaglia. *The Newman Lectures on Thermodynamics.* Singapore: Jenny Stanford Publishing (formerly Pan Stanford Publishing), 2019.
3. John Newman and Nitash P. Balsara. *Electrochemical Systems,* 4th Edition. Hoboken, New Jersey: John Wiley and Sons, 2020.

Section A:
Basic Transport Relations

Chapter 1

Conservation Laws and Transport Laws

Let us look briefly at heat transfer in solids. For simplicity, assume that the density ρ is constant.

The first law of thermodynamics says that

$$\begin{Bmatrix}\text{rate of accumulation}\\ \text{of internal energy}\end{Bmatrix} = \begin{Bmatrix}\text{net rate of heat}\\ \text{addition by conduction}\end{Bmatrix},$$

which may be expressed mathematically as

$$\frac{\partial}{\partial t}(\rho\hat{U}) = -\nabla\cdot\mathbf{q}.$$

From the definition of heat capacity,

$$\frac{\partial}{\partial t}(\rho\hat{U}) = \boxed{\rho\hat{C}_V\frac{\partial T}{\partial t} = -\nabla\cdot\mathbf{q}.}$$

It is also necessary to use Fourier's law of heat conduction:

$$\boxed{\mathbf{q} = -k\nabla T.}$$

Then

$$\rho\hat{C}_V\frac{\partial T}{\partial t} = \nabla\cdot(k\nabla T),$$

and for constant thermal conductivity k,

$$\rho\hat{C}_V\frac{\partial T}{\partial t} = k\nabla^2 T = k\left(\frac{\partial^2 T}{\partial x^2} + \frac{\partial^2 T}{\partial y^2} + \frac{\partial^2 T}{\partial z^2}\right).$$

The Newman Lectures on Transport Phenomena
John Newman and Vincent Battaglia

ISBN 978-981-4774-27-7 (Hardcover), 978-1-315-10829-2 (eBook)
www.jennystanford.com

These concepts provide the basis for the analysis of heat transfer in solids. They also illustrate a general problem. In the study of transport processes, we find a constant interplay of conservation laws and transport laws, such as the first law of thermodynamics and the Fourier law of heat conduction. We want to investigate in some detail the correct formulation of these laws for fluid mechanics, heat transfer, and mass transfer.

	Fluid mechanics	**Heat transfer**	**Mass transfer**
Conservation law	$F = ma$ (also conservation of mass)	First law of thermodynamics	Conservation of mass (by species)
Transport law	$\tau_{xy} = -\mu \frac{dv_x}{dy}$	$\mathbf{q} = -k\nabla T$	$\mathbf{J}_A^* = -c\mathcal{D}_{AB}\nabla x_A$

By way of a historical survey, we might note that:

1. In 1755, Euler wrote a differential form of the momentum balance for fluids, but he neglected viscous forces.
2. In 1822, Navier obtained the Navier–Stokes equation.
3. Fourier's law of heat conduction was stated by Biot (1804, 1816) and later by Fourier [1822 (1811)].
4. In 1855, Fick expressed a mass flux in terms of a concentration gradient.

Progress since this time has consisted of making the macroscopic theory more coherent, of giving the macroscopic theory a microscopic basis by means of statistical mechanics, of obtaining theoretical and experimental information about transport properties, of obtaining solutions to the equations for simple transport problems, and of empirical studies of more complicated systems.

The above comments should suggest the intimate connection of these studies with other fields of physical science, such as thermodynamics, chemical kinetics, and statistical mechanics.

Chapter 2

Fluid Mechanics

The principal result of this chapter is the basic equations of fluid mechanics:

Conservation of mass

$$\frac{\partial \rho}{\partial t} = -\nabla \cdot (\rho \mathbf{v}), \tag{2.1}$$

Conservation of momentum

$$\frac{\partial \rho \mathbf{v}}{\partial t} = -\nabla \cdot (\rho \mathbf{v}\, \mathbf{v}) - \nabla p - \nabla \cdot \boldsymbol{\tau} + \rho \mathbf{g}, \tag{2.2}$$

Momentum flux for a Newtonian fluid

$$\boldsymbol{\tau} = -\mu[\nabla \mathbf{v} + (\nabla \mathbf{v})^*] + \frac{2}{3}\mu \mathbf{I} \nabla \cdot \mathbf{v}. \tag{2.3}$$

Thus, there are two conservation laws and one transport law or flux relation.

2.1 Conservation of Mass

The differential expression 2.1 for conservation of mass can be obtained by equating the accumulation to the net input for a small, rectangular parallelepiped:

The Newman Lectures on Transport Phenomena
John Newman and Vincent Battaglia

ISBN 978-981-4774-27-7 (Hardcover), 978-1-315-10829-2 (eBook)
www.jennystanford.com

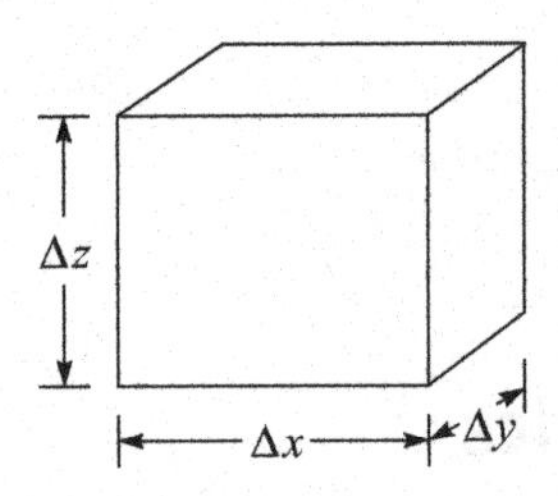

$$\frac{\partial}{\partial t}(\rho\Delta x\Delta y\Delta z)=\left(\rho v_x\big|_x-\rho v_x\big|_{x+\Delta x}\right)\Delta y\Delta z$$
$$+\left(\rho v_y\big|_y-\rho v_y\big|_{y+\Delta y}\right)\Delta x\Delta z$$
$$+\left(\rho v_z\big|_z-\rho v_z\big|_{z+\Delta z}\right)\Delta x\Delta y.$$

Divide by Δx, Δy, and Δz, and let each go to zero.

$$\frac{\partial\rho}{\partial t}=\lim_{\Delta x\to 0}\frac{\rho v_x\big|_x-\rho v_x\big|_{x+\Delta x}}{\Delta x}$$
$$+\lim_{\Delta y\to 0}\frac{\rho v_y\big|_y-\rho v_y\big|_{y+\Delta y}}{\Delta y}+\lim_{\Delta z\to 0}\frac{\rho v_z\big|_z-\rho v_z\big|_{z+\Delta z}}{\Delta z}$$
$$=-\frac{\partial\rho v_x}{\partial x}-\frac{\partial\rho v_y}{\partial y}-\frac{\partial\rho v_z}{\partial z}.$$

This is expressed more compactly in vector notation in Eq. 2.1. More elegant derivations involving an arbitrary volume element instead of the parallelepiped are sometimes employed, but the present procedure suffices to illustrate the point.

2.2 Conservation of Momentum

The law of conservation of momentum 2.2 is a differential expression of Newton's second law of motion. The component in the x-direction of such a force balance for a volume element is

$$\left\{\begin{matrix}\text{rate of accumulation}\\ \text{of } x\text{-momentum}\end{matrix}\right\}=\left\{\begin{matrix}\text{net rate of input}\\ \text{of } x\text{-momentum}\\ \text{by convection}\end{matrix}\right\}+\left\{\begin{matrix}\text{sum of forces}\\ \text{in the } x\text{-direction}\end{matrix}\right\}.$$

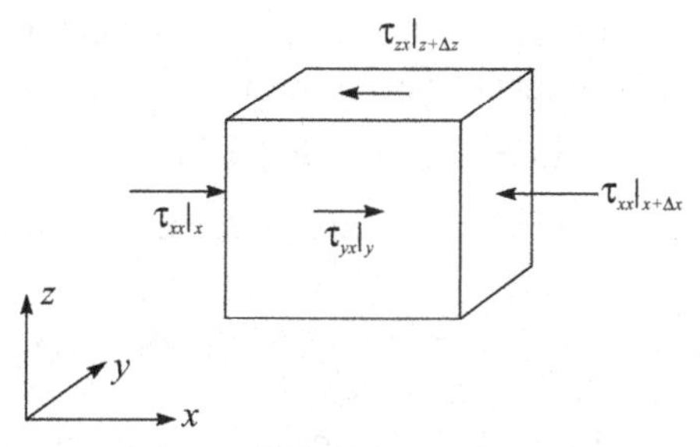

For a fluid, we want to include viscous forces, pressure forces, and gravitational forces. The momentum balance for our parallelepiped becomes

$$\begin{aligned}\frac{\partial}{\partial t}(\rho v_x \Delta x \Delta y \Delta z) &= \left(\rho v_x v_x \big|_x - \rho v_x v_x \big|_{x+\Delta x}\right)\Delta y \Delta z \\ &+\left(\rho v_x v_y \big|_y - \rho v_x v_y \big|_{y+\Delta y}\right)\Delta x \Delta z \\ &+\left(\rho v_x v_z \big|_z - \rho v_x v_z \big|_{z+\Delta z}\right)\Delta x \Delta y \\ &+\left(p\big|_x - p\big|_{x+\Delta x}\right)\Delta y \Delta z + \rho g_x \Delta x \Delta y \Delta z \\ &+\left(\tau_{xx}\big|_x - \tau_{xx}\big|_{x+\Delta x}\right)\Delta y \Delta z + \left(\tau_{yx}\big|_y - \tau_{yx}\big|_{y+\Delta y}\right)\Delta x \Delta z \\ &+\left(\tau_{zx}\big|_z - \tau_{zx}\big|_{z+\Delta z}\right)\Delta x \Delta y.\end{aligned}$$

Divide by Δx, Δy, and Δz, and let each approach zero.

$$\begin{aligned}\frac{\partial \rho v_x}{\partial t} &= -\frac{\partial \rho v_x v_x}{\partial x} - \frac{\partial \rho v_x v_y}{\partial y} - \frac{\partial \rho v_x v_z}{\partial z} - \frac{\partial p}{\partial x} \\ &+\rho g_x - \frac{\partial \tau_{xx}}{\partial x} - \frac{\partial \tau_{yx}}{\partial y} - \frac{\partial \tau_{zx}}{\partial z}.\end{aligned}$$

This is the x-component of Eq. 2.2.

Equations 2.1 and 2.2 are frequently written in an alternative form.

$$\frac{\partial \rho}{\partial t} = -\nabla \cdot (\rho \mathbf{v}) = -\rho \nabla \cdot \mathbf{v} - \mathbf{v} \cdot \nabla \rho.$$

$$\boxed{\frac{D\rho}{Dt} = \frac{\partial \rho}{\partial t} + \mathbf{v} \cdot \nabla \rho = -\rho \nabla \cdot \mathbf{v}.} \quad \text{conservation of mass} \qquad (2.4)$$

The operator D/Dt is defined as $\partial/\partial t + \mathbf{v}\cdot\nabla$.

The momentum equation can be manipulated in a similar fashion by using the relationship

$$\frac{\partial \rho \mathbf{v}}{\partial t} + \nabla \cdot (\rho \mathbf{v}\, \mathbf{v}) = \rho \frac{\partial \mathbf{v}}{\partial t} + \mathbf{v} \frac{\partial \rho}{\partial t} + \nabla \cdot (\rho \mathbf{v})\mathbf{v} + \rho \mathbf{v} \cdot \nabla \mathbf{v}.$$

Here two terms cancel by Eq. 2.1, and a vector identity for differentiation of products has been used

$$\nabla\cdot(\mathbf{a}\,\mathbf{b}) = (\nabla\cdot\mathbf{a})\mathbf{b} + \mathbf{a}\cdot\nabla\mathbf{b}$$

(see Problem A.2.) Thus, the momentum equation may be written as

$$\boxed{\rho\frac{D\mathbf{v}}{Dt} = \rho\left(\frac{\partial\mathbf{v}}{\partial t} + \mathbf{v}\cdot\nabla\mathbf{v}\right) = -\nabla p + \rho\mathbf{g} - \nabla\cdot\boldsymbol{\tau}.} \tag{2.5}$$

In this form, the similarity to Newton's second law of motion should be evident.

$$m\mathbf{a} = \mathbf{F}.$$

Here the force is composed of pressure forces, gravitational forces, and viscous forces.

2.3 Momentum Flux

It is usually necessary to say more about the viscous stress $\boldsymbol{\tau}$. The viscous stress is related to the rate of shear or rate of strain, that is, to velocity derivatives. A useful expression can be obtained on the basis of the Navier–Stokes stress assumptions:

1. The viscous stress is zero when the velocity is uniform ($\nabla\mathbf{v} = 0$).
2. The viscous stress depends linearly on the first velocity derivatives.
3. The fluid is isotropic (has no preferred directions).
4. $\boldsymbol{\tau}$ is symmetric. (This is not a physical assumption; see Problem 2.1.)

A straightforward, but tedious, investigation of the implications of these assumptions leads to

$$\boldsymbol{\tau} = -\mu[\nabla\mathbf{v} + (\nabla\mathbf{v})^*] + \left(\frac{2}{3}\mu - \kappa\right)\mathbf{I}\,\nabla\cdot\mathbf{v}$$

or

$$\tau_{ij} = -\mu\left(\frac{\partial v_i}{\partial x_j} + \frac{\partial v_j}{\partial x_i}\right) + \left(\frac{2}{3}\mu - \kappa\right)\delta_{ij}\nabla\cdot\mathbf{v}$$

where $\delta_{ij} = 1$ when $i = j$ and $\delta_{ij} = 0$ when $i \neq j$. More specifically, the diagonal elements look like

$$\tau_{xx} = -2\mu\frac{\partial v_x}{\partial x} + \left(\frac{2}{3}\mu - \kappa\right)\nabla\cdot\mathbf{v},$$

while off-diagonal elements look like

$$\tau_{xy} = \tau_{yx} = -\mu\left(\frac{\partial v_x}{\partial y} + \frac{\partial v_y}{\partial x}\right).$$

In these equations, we have introduced two *transport properties*, μ and κ. These are called the viscosity and bulk viscosity, respectively, and are functions of the temperature, pressure, and composition of the fluid. The bulk viscosity is sufficiently small and sufficiently difficult to measure that we shall usually neglect it and adopt Eq. 2.3 as the expression of the viscous momentum flux for a *Newtonian fluid.*

2.4 Assumptions

An important assumption that we shall make throughout most of our work is that thermodynamic relations can be applied locally. Thus, we define the temperature at a point in a heat-conducting medium even though thermodynamics applies only to equilibrium situations.

The second assumption is that we can use continuum mechanics and do not need to account for the discrete nature of matter. This is usually a valid and fruitful approach. It has also been assumed that mass and momentum are conserved, and any external electromagnetic forces have been neglected. Actually there are few serious assumptions until you begin to specify $\boldsymbol{\tau}$. For example, Eqs. 2.1 and 2.2 apply to an elastic solid, which by no means satisfy Eq. 2.3.

The Navier–Stokes stress assumption does not apply to plastics (nonnewtonian fluids) or rarefied gases or to cases where the stresses are extremely large. In these cases, the stress may be a nonlinear function of the velocity derivatives, and for a solid, the stress does not need to be zero when the velocity is uniform. Fluids with long oriented polymer molecules may not even be isotropic. Finally, there is no cogent reason why the stress at one point should not depend on conditions at other points in the fluid or on the history of the fluid motion.

Despite the importance of nonnewtonian fluids, we shall mainly confine our attention to Newtonian fluids.

Problems

2.1 Problem 3L of Ref. [1]. The stress tensor must be symmetric even for nonnewtonian fluids.

2.2 Principal axes for stress.

The existence of "normal viscous stresses" frequently confuses the readers. In this regard, it might be helpful to point out that any symmetric tensor can be made diagonal merely by rotating the coordinate axes. In the new coordinate system, there are normal viscous stresses but no tangential viscous stresses (and the coordinate system is said to be referred to the principal axes of the tensor).

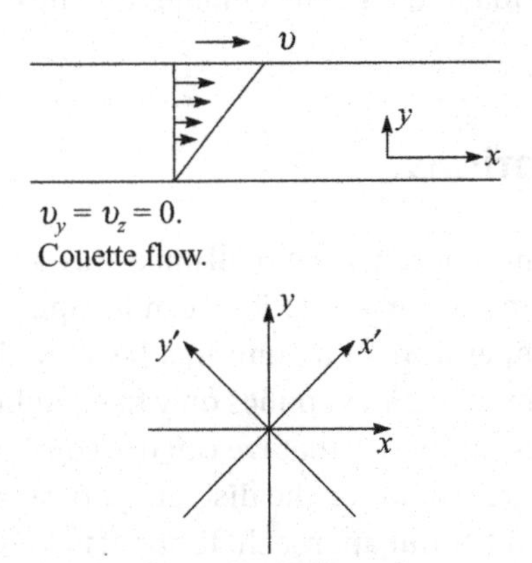

$v_y = v_z = 0$.
Couette flow.

Rotated coordinates.

As an example, consider Couette flow and let the only non-vanishing velocity derivative be dv_x/dy. Show that in a coordinate system rotated by 45° about the z-axis, there are no tangential stresses.

Our intuition is led astray by the fact that the normal stress on a solid surface in an incompressible fluid is zero. This merely tells us that the solid surface is not perpendicular to the principal axis of stress, not that normal stresses cannot exist.

Reference

1. R. Byron Bird, Warren E. Stewart, and Edwin N. Lightfoot. *Transport Phenomena*. New York: John Wiley & Sons, 1960.

Chapter 3

Microscopic Interpretation of the Momentum Flux

The equation of conservation of momentum 2.2 can be written as

$$\frac{\partial \rho \mathbf{v}}{\partial t} = -\nabla \cdot \boldsymbol{\phi} + \rho \mathbf{g} \tag{3.1}$$

where

$$\boldsymbol{\phi} = \rho \mathbf{v}\,\mathbf{v} + p\mathbf{I} + \boldsymbol{\tau}. \tag{3.2}$$

$\boldsymbol{\phi}$ can thus be regarded as the momentum flux in the fluid, and Eq. 3.1 is analogous to the equation of conservation of mass 2.1. In Eq. 3.1, there is an additional term $\rho\mathbf{g}$ representing an interaction of the fluid with a gravitational field.

On a microscopic scale, $\boldsymbol{\phi}$ can be regarded as an average over the momenta carried by the individual fluid particles. For a single component fluid,

$$\boldsymbol{\phi} = \int m\,\mathbf{v}\,\mathbf{v}\,f(\mathbf{v})\,d^3v \;+ \text{intermolecular attraction.} \tag{3.3}$$

Here $m\mathbf{v}$ is the momentum of a particle moving with a velocity $\mathbf{v}$, and this momentum is, of course, transported with the velocity $\mathbf{v}$, giving a contribution of $m\mathbf{v}\,\mathbf{v}$ to the momentum flux. All particles do not move with the same velocity, but the *velocity distribution function* $f(\mathbf{v})$ expresses the random velocities of the particles. This is defined in such a way that the mass density is given by

$$\rho = \int m\,f(\mathbf{v})\,d^3v \tag{3.4}$$

and the average velocity $\mathbf{v}$ is given by

The Newman Lectures on Transport Phenomena
John Newman and Vincent Battaglia

ISBN 978-981-4774-27-7 (Hardcover), 978-1-315-10829-2 (eBook)
www.jennystanford.com

$$\rho \mathbf{v} = \int m\mathbf{v}\, f(\mathbf{v})\, d^3v. \tag{3.5}$$

Equation 3.4 merely expresses the normalization of the velocity distribution function. We can see that Eq. 3.5 is written in a form that indicates that $\rho\mathbf{v}$ is the average mass flux. This, of course, was what we had in mind when we wrote Eq. 2.1 for conservation of mass. The term "intermolecular attractions" in Eq. 3.3 refers to the fact that the only contribution to the momentum flux does not come from the moving molecules. Intermolecular forces become relatively more important in liquids, while transport of molecular momentum dominates in gases.

We also see from Eq. 3.5 that $\rho\mathbf{v}$ is the average momentum density (per unit volume). Hence, it enters on the left in the equation of conservation of momentum 3.1. The momentum flux 3.3 can be broken down since

$$\begin{aligned}\int m\mathbf{v}\mathbf{v} f(\mathbf{v})\, d^3v &= \int m(\mathbf{v}-\mathbf{v}+\mathbf{v})(\mathbf{v}-\mathbf{v}+\mathbf{v}) f(\mathbf{v}) d^3v \\ &= \int m(\mathbf{v}-\mathbf{v})(\mathbf{v}-\mathbf{v}) f(\mathbf{v}) d^3v + \mathbf{v}\mathbf{v}\int m f(\mathbf{v})\, d^3v\end{aligned}$$

and since

$$\int (\mathbf{v}-\mathbf{v}) f(\mathbf{v}) d^3v = 0.$$

Hence

$$\boldsymbol{\phi} = \rho\mathbf{v}\mathbf{v} + \int m(\mathbf{v}-\mathbf{v})(\mathbf{v}-\mathbf{v}) f(\mathbf{v}) d^3v + \text{imf}^{\dagger}.$$

Comparison with Eq. 3.2 shows that

$$\boldsymbol{\tau} + p\mathbf{I} = \int m(\mathbf{v}-\mathbf{v})(\mathbf{v}-\mathbf{v}) f(\mathbf{v}) d^3v + \text{imf}, \tag{3.6}$$

which thus represents the momentum flux due to deviations of individual particles from the average velocity. Equation 3.6 suggests the symmetric nature of $\boldsymbol{\tau}$. There is, however, nothing in this chapter to indicate that the fluid might be Newtonian. Additional justification of Eq. 2.3 can, however, be obtained in the kinetic theory of gases by an extension of these concepts but with additional assumptions as to the interaction of the molecules.

These considerations of microscopic nature are somewhat parenthetical and are not essential to an understanding of the bulk of the material in this book.

† imf stands for intermolecular force terms. These are physically important, but they do not help to illustrate the point here.

Chapter 4

Heat Transfer in a Pure Fluid

It is interesting to summarize occasionally the mathematical description of a problem and to count the equations and variables.

Equations for a pure fluid		Variables	
$\frac{\partial \rho}{\partial t} = -\nabla \cdot (\rho \mathbf{v})$	1	ρ	1
		$\mathbf{v}$	3
$\frac{\partial \rho \mathbf{v}}{\partial t} = -\nabla \cdot (\rho \mathbf{v}\, \mathbf{v}) - \nabla p - \nabla \cdot \boldsymbol{\tau} + \rho \mathbf{g}$	3	p	1
		$\boldsymbol{\tau}$	9
$\boldsymbol{\tau} = -\mu\left(\nabla \mathbf{v} + (\nabla \mathbf{v})^*\right) + \frac{2}{3}\mu \mathbf{I} \nabla \cdot \mathbf{v}$	9	μ	1
$p = p(\rho, T)$	1	T	$\frac{1}{16}$
$\mu = \mu(\rho, T)$	$\frac{1}{15}$	(assuming $\mathbf{g}$ to be given)	

The Newman Lectures on Transport Phenomena
John Newman and Vincent Battaglia

ISBN 978-981-4774-27-7 (Hardcover), 978-1-315-10829-2 (eBook)
www.jennystanford.com

As long as the temperature is constant, we can solve the problem. However, viscous flow causes a temperature rise, and rather than dismissing this qualitatively, let us look at the energy equation. This can be derived by applying the first law of thermodynamics to a volume element.

rate of accumulation of internal and kinetic energy
= net rate of internal and kinetic energy in by convection
+ net rate of heat addition by conduction
− net rate of work done by system on surroundings.

The details of the derivation are very similar to those for the equations of continuity and motion in Chapter 2, and they will not be reproduced here. The result is

$$\frac{\partial}{\partial t}\left(\rho\hat{U}+\frac{1}{2}\rho v^2\right)=-\nabla\cdot\left[\left(\rho\hat{U}+\frac{1}{2}\rho v^2\right)\mathbf{v}\right]-\nabla\cdot\mathbf{q} \tag{4.1}$$
$$+\rho\mathbf{v}\cdot\mathbf{g}-\nabla\cdot(p\mathbf{v})-\nabla\cdot(\boldsymbol{\tau}\cdot\mathbf{v}).$$

The work done against gravitational forces, pressure forces, and viscous forces is included, but the absorption and emission of radiant energy and other interactions with electromagnetic fields have been ignored.

An apparent simplification of the complete energy equations can be achieved with the aid of the equation of motion. The dot product of $\mathbf{v}$ with the equation of motion gives the "equation of mechanical energy."

$$\rho\mathbf{v}\cdot\left(\frac{\partial\mathbf{v}}{\partial t}+\mathbf{v}\cdot\nabla\mathbf{v}\right)=-\mathbf{v}\cdot\nabla p-\mathbf{v}\cdot\nabla\cdot\boldsymbol{\tau}+\rho\mathbf{g}\cdot\mathbf{v}.$$

The relations

$$\frac{\partial\rho v^2}{\partial t}+\nabla\cdot\left(\rho v^2\mathbf{v}\right)=2\rho\mathbf{v}\cdot\left[\frac{\partial\mathbf{v}}{\partial t}+\mathbf{v}\cdot\nabla\mathbf{v}\right]$$

$$\nabla\cdot(p\mathbf{v})=p\nabla\cdot\mathbf{v}+\mathbf{v}\cdot\nabla p$$

and

$$\nabla\cdot(\boldsymbol{\tau}\cdot\mathbf{v})=\boldsymbol{\tau}:\nabla\mathbf{v}+\mathbf{v}\cdot\nabla\cdot\boldsymbol{\tau}$$

allow this to be expressed in the more common form

$$\frac{\partial}{\partial t}\left(\frac{1}{2}\rho v^2\right)+\nabla\cdot\left(\frac{1}{2}\rho v^2\mathbf{v}\right)=p\nabla\cdot\mathbf{v}-\nabla\cdot(p\mathbf{v})+\boldsymbol{\tau}:\nabla\mathbf{v}-\nabla\cdot(\boldsymbol{\tau}\cdot\mathbf{v})+\rho\mathbf{v}\cdot\mathbf{g}.$$

Subtraction from the energy equation yields the "thermal energy equation:"

$$\boxed{\frac{\partial \rho \hat{U}}{\partial t} = -\nabla \cdot (\rho \hat{U} \mathbf{v}) - \nabla \cdot \mathbf{q} - p \nabla \cdot \mathbf{v} - \boldsymbol{\tau} : \nabla \mathbf{v}.} \tag{4.2}$$

This procedure is known as the "splitting" of the energy equation.

Let us return to the collection of equations for pure fluids. Here an entirely equivalent form of the thermal energy equation is

$$\rho \hat{C}_V \frac{DT}{Dt} = -\nabla \cdot \mathbf{q} - T \left(\frac{\partial p}{\partial T} \right)_{\rho} \nabla \cdot \mathbf{v} - \boldsymbol{\tau} : \nabla \mathbf{v}, \tag{4.3}$$

as can be seen by means of the thermodynamic relation

$$d\hat{U} = \left[-p + T \left(\frac{\partial p}{\partial T} \right)_{\hat{V}} \right] d\hat{V} + \hat{C}_V dT.$$

For a pure fluid, the heat flux **q** can be expressed by Fourier's law of heat conduction:

$$\mathbf{q} = -k \nabla T. \tag{4.4}$$

The extension of our list of equations and variables then reads:

Equations		**Variables**	
equation 4.3	1	$\hat{C}_V$	1
$\mathbf{q} = -k \nabla T$	3	$\mathbf{q}$	3
$k = k(\rho, T)$	1	k	1
			21
$\hat{C}_V = \hat{C}_V(\rho, T)$	1		
	21		

We now have enough equations to treat pure fluids. Fluid mixtures will, of course, be more complicated.

Let us also consider a microscopic interpretation of the energy equation 4.1, which can be written as

$$\frac{\partial}{\partial t} \left(\rho \hat{U} + \frac{1}{2} \rho v^2 \right) = -\nabla \cdot \mathbf{e} + \rho \mathbf{v} \cdot \mathbf{g} \tag{4.5}$$

where the energy flux is

$$\mathbf{e} = \left(\rho \hat{U} + \frac{1}{2} \rho v^2 \right) \mathbf{v} + \mathbf{q} + p \mathbf{v} + \boldsymbol{\tau} \cdot \mathbf{v}. \tag{4.6}$$

We should expect to express the total energy density for a pure fluid as

$$\rho\hat{K} = \int \left[\frac{1}{2} m v^2 + \epsilon\right] f(\mathbf{v}) d^3 v + \text{ima}, \tag{4.7}$$

where $\epsilon(\mathbf{r}, \mathbf{v}, t)$ is the internal energy of the molecules at $\mathbf{r}$, $\mathbf{v}$, t averaged over the rotational and vibrational degrees of freedom. Again, "ima" denotes the fact that there is a potential energy associated with intermolecular attractions.

By identifying the macroscopic kinetic energy,

$$\rho\hat{K} = \frac{1}{2}\rho \upsilon^2 + \rho\hat{U} \tag{4.8}$$

we find the internal energy to be given by

$$\rho\hat{U} = \int \left[\frac{1}{2} m(\mathbf{v} - \mathbf{v})^2 + \epsilon\right] f(\mathbf{v}) d^3 v + \text{ima}. \tag{4.9}$$

The energy flux would be

$$\mathbf{e} = \int \left[\frac{1}{2} m v^2 + \epsilon\right] \mathbf{v}\, f(\mathbf{v}) d^3 v + \text{ima}. \tag{4.10}$$

If we break this down to

$$\mathbf{e} = \rho\hat{K}\mathbf{v} + (\boldsymbol{\tau} + p\mathbf{I}) \cdot \mathbf{v} + \mathbf{q}$$

then the residue is

$$\mathbf{q} = \int \left[\frac{1}{2} m(\mathbf{v} - \mathbf{v})^2 + \epsilon\right] (\mathbf{v} - \mathbf{v})\, f(\mathbf{v}) d^3 v + \text{ima}. \tag{4.11}$$

The internal energy $\rho\hat{U}$ is composed of the random kinetic energy of the molecules as distinct from the coherent kinetic energy $\frac{1}{2}\rho\upsilon^2$. The rotational and vibrational energy of the molecules represents a complication that did not appear when we looked at the momentum relations.

The equation

$$\frac{\partial \rho\hat{K}}{\partial t} = -\nabla \cdot \mathbf{e} + \rho\mathbf{v} \cdot \mathbf{g}$$

represents the rate of change of total energy (Eq. 4.7) due to the energy flux (Eq. 4.10) and interaction with the gravitational field

$$\rho\mathbf{v} \cdot \mathbf{g} = \int m\mathbf{v} \cdot \mathbf{g} f(\mathbf{v}) d^3 v.$$

The energy flux decomposes quite naturally into a macroscopic convective energy flux $\rho\hat{K}\mathbf{v}$, work done against the surface forces

$p\mathbf{I}+\boldsymbol{\tau}$ identified in our earlier inspection of the momentum equation, and a residual molecular energy flux $\mathbf{q}$ with respect to the average velocity. The expression for $\mathbf{q}$ says that molecules moving faster than average carry more than average energy so that diffusion of high-speed molecules from regions of high temperature to regions of low temperature creates an energy flux with respect to the average velocity.

Problems

4.1 (a) For condensed phases, the heat capacity at constant volume and the heat capacity at constant pressure are supposed to be nearly the same. From the definitions

$$C_V=\left(\frac{\partial U}{\partial T}\right)_V \quad \text{and} \quad C_p=\left(\frac{\partial H}{\partial T}\right)_p,$$

show that the two are related by

$$C_p=C_V+T\left(\frac{\partial p}{\partial T}\right)_V\left(\frac{\partial V}{\partial T}\right)_p.$$

Calculate the percent difference between C_V and C_p for an ideal diatomic gas and for water, for which the compressibility and the coefficient of thermal expansion at 20°C are

$$\kappa=-\frac{1}{V}\left(\frac{\partial V}{\partial p}\right)_T=4.58\times10^{-11}\,\frac{\text{cm}^2}{\text{dyne}}$$

$$\text{and } \alpha=\frac{1}{V}\left(\frac{\partial V}{\partial T}\right)_p=0.207\times10^{-3}\,\text{K}^{-1},$$

and the heat capacity is

$$\hat{C}_p=0.99883\ \text{cal/g-K}=4.1819\ \text{J/g-K}.$$

(b) The thermal energy equation for a pure fluid is

$$\rho\hat{C}_V\frac{DT}{Dt}=-\nabla\cdot\mathbf{q}-T\left(\frac{\partial p}{\partial T}\right)_\rho\nabla\cdot\mathbf{v}-\boldsymbol{\tau}:\nabla\mathbf{v}.$$

For a condensed phase, we should like to treat the density as a constant, $\nabla\cdot\mathbf{v}$ = 0, but we are bothered that the coefficient $(\partial p/\partial T)_\rho$ is large. Show that the above form of the thermal energy equation is completely equivalent to the equation

$$\rho \hat{C}_p \frac{DT}{Dt} = -\nabla \cdot \mathbf{q} - \boldsymbol{\tau} : \nabla \mathbf{v} + \frac{T}{\hat{V}} \left(\frac{\partial \hat{V}}{\partial T} \right)_p \frac{Dp}{Dt}.$$

(c) For water, a change in pressure of 1 atm in the last term of this equation corresponds to what temperature change in the first term?

Chapter 5

Concentrations and Velocities in Mixtures

The average velocity $\mathbf{v}$ for a pure fluid was expressed by Eq. 3.5. Let us similarly define an average velocity for each component of a mixture:

$$\mathbf{v}_i = \frac{1}{\rho_i}\int m_i \mathbf{v}\, f_i(\mathbf{v})\, d^3v, \tag{5.1}$$

where the normalization of the velocity distribution functions f_i is specified by the expression for the mass density of the components:

$$\rho_i = \int m_i f_i(\mathbf{v}) d^3v. \tag{5.2}$$

It should be clear from our definition that $\mathbf{v}_i$ is an average velocity for a component of the mixture and does not represent the velocity of individual molecules.

The total density of the mixture is

$$\rho = \sum_i \rho_i = \sum_i \int m_i f_i(\mathbf{v})\, d^3v. \tag{5.3}$$

The momentum density of a component is

$$\int m_i \mathbf{v}\, f_i(\mathbf{v})\, d^3v = \rho_i \mathbf{v}_i$$

and the total momentum density of the fluid is

$$\sum_i \int m_i \mathbf{v}\, f_i(\mathbf{v})\, d^3v = \sum_i \rho_i \mathbf{v}_i.$$

The Newman Lectures on Transport Phenomena
John Newman and Vincent Battaglia

ISBN 978-981-4774-27-7 (Hardcover), 978-1-315-10829-2 (eBook)
www.jennystanford.com

Since the momentum density in the equation of motion 2.2 is $\rho\mathbf{v}$ and since we should like this equation to remain valid, we define $\mathbf{v}$ as the "mass-average velocity"

$$\mathbf{v} = \frac{1}{\rho}\sum_i \rho_i \mathbf{v}_i. \tag{5.4}$$

The other terms ∇p and $\rho\mathbf{g}$ in the equation of motion still have meaning and are valid as long as the external force per unit mass $\mathbf{g}$ is the same for all species, which we shall assume. As usual, the term $-\nabla\cdot\boldsymbol{\tau}$ accounts for everything else that should go into the momentum balance. Actually a mixture can still be Newtonian,

$$\boldsymbol{\tau} = -\mu[\nabla\mathbf{v} + (\nabla\mathbf{v})^*] + \frac{2}{3}\mu\mathbf{I}\,\nabla\cdot\mathbf{v},$$

but now μ is a function of temperature, pressure, and composition.

The total mass flux in the mixture is

$$\sum_i \int m_i \mathbf{v} f_i(\mathbf{v}) d^3v = \sum_i \rho_i \mathbf{v}_i = \rho\mathbf{v},$$

which is the same as the momentum density. Hence the continuity equation

$$\frac{\partial\rho}{\partial t} = -\nabla\cdot(\rho\mathbf{v})$$

still has its originally intended meaning.

The preceding discussion illustrates the convenience of the mass-average velocity. This velocity allows the equation of continuity and the equation of motion to describe a mixture. However, chemists frequently like to work with moles. The number density of component i is

$$\int f_i(\mathbf{v})\, d^3v.$$

We define the (molar) concentration of a species as

$$c_i = \frac{1}{L}\int f_i(\mathbf{v})\, d^3v.$$

The total concentration is

$$c = \frac{1}{L}\sum_i \int f_i(\mathbf{v}) d^3v = \sum_i c_i.$$

The molar average velocity is defined as the average number of particles crossing a unit area per unit time divided by the total number density:

$$\mathbf{v}^* = \frac{\sum_i \int \mathbf{v} f_i(\mathbf{v}) d^3v}{\sum_i \int f_i(\mathbf{v})\, d^3v} = \frac{1}{c} \sum_i c_i \mathbf{v}_i.$$

The convenience of these definitions is related to the fact that the pressure in a dilute gas mixture is more simply related to c than to ρ:

$$p = cRT,$$

or, in other words, chemists do not like the gas constant R to depend on the molar mass.

Chapter 6

Material Balances and Diffusion

The molar flux of species i is

$$\mathbf{N}_i = c_i \mathbf{v}_i = \frac{1}{L}\int \mathbf{v}\, f_i(\mathbf{v}) d^3 v, \tag{6.1}$$

and the mass flux is

$$\mathbf{n}_i = \rho_i \mathbf{v}_i = \int m_i \mathbf{v}\, f_i(\mathbf{v}) d^3 v, \tag{6.2}$$

where $\rho_i = M_i c_i$.

A differential material balance

accumulation = net input + production

for each species takes the form

$$\frac{\partial c_i}{\partial t} = -\nabla \cdot (c_i \mathbf{v}_i) + R_i \quad \text{or} \quad \frac{\partial \rho_i}{\partial t} = -\nabla \cdot (\rho_i \mathbf{v}_i) + r_i. \tag{6.3}$$

The second equation is the first equation multiplied by M_i. The rate of production of species i by chemical (or nuclear) reactions is denoted by R_i when expressed in moles per unit volume per unit time and by r_i when expressed in grams. Since mass is conserved, $\sum_i r_i = 0$. The problem of expressing R_i as a function of temperature and composition is the subject of applied chemical kinetics.

Now it is necessary to consider in more detail the molar flux $\mathbf{N}_i = c_i \mathbf{v}_i$ and the mass flux $\mathbf{n}_i = \rho_i \mathbf{v}_i$. Immediately, two contributions to the flux come to mind: bulk motion of the fluid as a whole and

The Newman Lectures on Transport Phenomena
John Newman and Vincent Battaglia

ISBN 978-981-4774-27-7 (Hardcover), 978-1-315-10829-2 (eBook)
www.jennystanford.com

diffusion of the species relative to each other due to concentration gradients. Clearly, we can describe the bulk motion of the fluid by means of some average velocity. Then diffusion represents a correction to this average motion (or a deviation from the average motion)

$$\mathbf{n}_i = \rho_i \mathbf{v} + \mathbf{j}_i, \quad \mathbf{N}_i = c_i \mathbf{v} + \mathbf{J}_i.$$

Of course, we could use the molar average velocity just as well:

$$\mathbf{n}_i = \rho_i \mathbf{v}^* + \mathbf{j}_i^*, \quad \mathbf{N}_i = c_i \mathbf{v}^* + \mathbf{J}_i^*.$$

From these definitions,

$$\sum_i \mathbf{j}_i = 0. \quad \sum_i \mathbf{J}_i^* = 0.$$

Now that we have accounted for the bulk motion, we are free to describe the diffusion flux $\mathbf{j}_i$ (or $\mathbf{J}_i$ or $\mathbf{j}_i^*$ or $\mathbf{J}_i^*$). If we use the fluxes $\mathbf{J}_i^*$, we can consider ourselves to be moving so that the molar average velocity is zero. The values of $\mathbf{J}_i^*$ must, therefore, be due to concentration gradients.

For a binary system containing species A and B, Fick's law of diffusion becomes

$$\mathbf{J}_A^* = -c\mathcal{D}_{AB}\nabla x_A. \tag{6.4}$$

Here x_A is the mole fraction of species A ($x_A = c_A/c$), and $\mathcal{D}_{AB}$ is the diffusion coefficient of species A in species B.

$$\mathbf{J}_B^* = -\mathbf{J}_A^* = -c\mathcal{D}_{AB}\nabla x_A = -c\mathcal{D}_{AB}\nabla x_B.$$

Hence, $\mathcal{D}_{AB} = \mathcal{D}_{BA}$.

Only one statement of the diffusion law is necessary. From the interrelationships among the fluxes, one can use molar or mass diffusion fluxes with respect to mass or molar average velocity, and these diffusion fluxes can be expressed in terms of a driving force of mole fraction gradient, concentration gradient, mass fraction gradient, etc. For example, Fick's law can be expressed as

$$\mathbf{j}_A = -\rho\mathcal{D}_{AB}\nabla\omega_A \tag{6.5}$$

where ω_A is the mass fraction of species A ($\omega_A = \rho_A/\rho$).

It should be noted that only one new transport coefficient, $\mathcal{D}_{AB}$, has been defined. Equation 6.5 can be derived from Eq. 6.4, and $\mathcal{D}_{AB}$ is the same quantity in the two equations. It is, of course, possible to define other diffusion coefficients, but for a binary system, there is only one independent diffusion coefficient that needs to be measured and tabulated. Any others can be calculated.

The expressions 6.4 and 6.5 of Fick's law will be refined in Chapter 8 by a consideration of pressure diffusion and thermal diffusion.

Problem .

6.1 Problem 17B.4 in Ref. [1], p. 541.

Reference

1. R. Byron Bird, Warren E. Stewart, and Edwin N. Lightfoot. *Transport Phenomena*, 2nd Edition. New York: John Wiley & Sons, 2002.

Chapter 7

Relaxation Time for Diffusion

The differential material balance for a species is

$$\frac{\partial c_i}{\partial t} = -\nabla \cdot (c_i \mathbf{v}_i) + R_i \qquad \text{(which is Eq. 6.3)},$$

and Fick's law of diffusion for a binary system can be expressed as

$$\mathbf{J}_A^* = -c\mathcal{D}_{AB} \nabla x_A = \frac{c_A c_B}{c}(\mathbf{v}_A - \mathbf{v}_B) \qquad \text{(which is Eq. 6.4)}.$$

The usual calculation procedure would be to calculate the mass-average velocity $\mathbf{v} = \frac{1}{\rho}\sum_i \rho_i \mathbf{v}_i$ with the equation of motion and then to use the diffusion equations to determine deviations of the species velocities from the mass-average velocity. The question may be raised: Shouldn't a separate momentum equation be written for each species?

Perhaps we should write a separate momentum equation. However, we assume that the relaxation time for the establishment of diffusion velocities is very small, and in this case the above procedure is satisfactory.

Let us illustrate these concepts for the motion of electrons in copper. Newton's second law of motion for the electrons is

$$\mathbf{F} = m_e \mathbf{a} = m_e \frac{d\mathbf{v}_e}{dt}, \tag{7.1}$$

The Newman Lectures on Transport Phenomena
John Newman and Vincent Battaglia

ISBN 978-981-4774-27-7 (Hardcover), 978-1-315-10829-2 (eBook)
www.jennystanford.com

and the force on the electrons may be expressed as

$$\mathbf{F} = -e_e \nabla \Phi - K(\mathbf{v}_e - \mathbf{v}_x). \tag{7.2}$$

The first term here is due to the electric field $-\nabla\Phi$, and the second term represents the resultant force from collisions with the crystal lattice. This last average force is assumed to be proportional to the difference in the average velocity of the electrons and the crystal lattice so that K is a constant.

The solution to these equations for an initial condition of $\mathbf{v}_e = \mathbf{v}_x$ at $t = 0$ is

$$\mathbf{v}_e = \mathbf{v}_x - \frac{e_e \nabla \Phi}{K}\left(1 - e^{-Kt/m_e}\right). \tag{7.3}$$

The steady-state solution corresponds to Ohm's law.

$$\mathbf{i} = e_e L c_e (\mathbf{v}_e - \mathbf{v}_x) = -\sigma \nabla \Phi = -\frac{e^2 L c_e \nabla \Phi}{K}$$

so that K can be related to the electrical conductivity σ

$$K = \frac{e^2 L c_e}{\sigma}.$$

On the other hand, the relaxation time

$$\tau = \frac{m_e}{K}$$

gives an idea of how long it takes for the steady velocity to be established.

For copper,

$$\sigma = \frac{1}{1.692 \times 10^{-6}} (\Omega\,\text{cm})^{-1} \text{ at } 20°\text{C}$$

$$\rho = 8.89\,\text{g/cm}^3$$

$$M = 63.54\,\text{g/mol}.$$

Hence, since $c_e = \rho/M$ (assuming one electron per atom), the relaxation time for copper is

$$\tau = \frac{m_e}{K} = \frac{m_e \sigma}{e^2 L c_e} = \frac{m_e \sigma M}{e^2 \rho L} = \frac{M^2 \sigma}{\rho F^2} \frac{m_e}{m_{\text{Cu}}}$$

$$= \frac{63.54/1830}{1.692 \times 10^{-6} \times 8.89 \times (0.965)^2 \times 10^{10} \times 10^7}\,\text{s}$$

$$= 2.45 \times 10^{-14}\,\text{s}.$$

Thus, Eq. 6.3 can be written as

$$\mathbf{v}_e - \mathbf{v}_x = -\frac{e\nabla\Phi}{K}\left(1 - e^{-t/\tau}\right)$$

where $\tau = 2.45 \times 10^{-14}$ s. If we are unable to observe changes in such a short time, we might as well approximate Eq. 6.3 by

$$\mathbf{v}_e - \mathbf{v}_x = -\frac{e\nabla\Phi}{K},$$

which is equivalent to the diffusion equation 6.4.

Chapter 8

Multicomponent Diffusion

For the motion of electrons in copper, we wrote the equation

$$\mathbf{F} = m_e \frac{d\mathbf{v}_e}{dt} = \underbrace{-e_e \nabla \Phi}_{\text{external force}} - \underbrace{K(\mathbf{v}_e - \mathbf{v}_x)}_{\text{interaction or drag force}} \approx 0.$$

In the extension to multicomponent diffusion, let the "external" force per unit volume on species i be denoted by

$$-\mathbf{d}_i \approx -c_i \nabla \mu_i. \tag{8.1}$$

Here we are using the gradient of chemical potential as a driving force for diffusion. This is opposed by the resistive forces due to the other species present. Let

$$\boldsymbol{\psi}_i^j = K_{ij}(\mathbf{v}_j - \mathbf{v}_i) \tag{8.2}$$

be the force exerted *on* species i by species j (per unit volume). This is a force due to the relative motion of the two species. Clearly, by Newton's third law of motion,

$$K_{ij} = K_{ji}. \tag{8.3}$$

If we ignore the relaxation time for acceleration, the total force exerted on species i can be equated to zero.

$$\mathbf{F} = -\mathbf{d}_i + \sum_j \boldsymbol{\psi}_i^j = -c_i \nabla \mu_i + \sum_j K_{ij}(\mathbf{v}_j - \mathbf{v}_i) = 0.$$

The Newman Lectures on Transport Phenomena
John Newman and Vincent Battaglia

ISBN 978-981-4774-27-7 (Hardcover), 978-1-315-10829-2 (eBook)
www.jennystanford.com

$$c_i \nabla \mu_i = \sum_j K_{ij}(\mathbf{v}_j - \mathbf{v}_i). \tag{8.4}$$

This is the equation of multicomponent diffusion in terms of friction coefficients K_{ij}. To this we can add embellishments.

First we define different transport coefficients by

$$K_{ij} = \frac{c_i c_j}{c\mathcal{D}_{ij}} RT, \quad \mathcal{D}_{ij} = \mathcal{D}_{ji}. \tag{8.5}$$

One interaction coefficient $\mathcal{D}_{ij}$ is defined for each *pair* of species. This amounts to $\frac{1}{2}n(n-1)$ coefficients, where n is the number of species present. Since the frequency of binary collisions between molecules of types i and j would be proportional to $c_i c_j$, most of the concentration dependence of K_{ij} is accounted for by Eq. 8.5, and the $\mathcal{D}_{ij}$ coefficients are relatively constant. Furthermore, for dilute gas mixtures, the multicomponent diffusion equation reduces to the Stefan–Maxwell equation

$$\nabla x_i = \sum_j \frac{x_i x_j}{\mathcal{D}_{ij}} (\mathbf{v}_j - \mathbf{v}_i). \tag{8.6}$$

The $\mathcal{D}_{ij}$ can then be conveniently measured in binary mixtures. These transport properties are called diffusion coefficients and have dimensions of cm^2/s.

Equation 8.4 summed over species yields

$$\sum_i c_i \nabla \mu_i = \sum_i \sum_j K_{ij}(\mathbf{v}_j - \mathbf{v}_i) = 0.$$

The right side is equal to zero by Newton's third law of motion (see Eq. 8.3). At constant temperature and pressure, the left side is also zero by the Gibbs–Duhem equation. Thus, we see that there are only n - 1 independent diffusion equations. When the temperature and pressure are not constant, the driving force should be modified to read

$$\mathbf{d}_i = c_i \left(\nabla \mu_i + \bar{S}_i \nabla T - \frac{M_i}{\rho} \nabla p \right). \tag{8.7}$$

Then $\sum_i \mathbf{d}_i = 0$ even when temperature and pressure vary.

The final multicomponent diffusion equation reads

$$\mathbf{d}_i = RT\sum_j \frac{c_i c_j}{c\mathcal{D}_{ij}}\left[\mathbf{v}_j - \mathbf{v}_i + \left(\frac{D_j^T}{\rho_j} - \frac{D_i^T}{\rho_i}\right)\nabla \ln T\right]. \tag{8.8}$$

The driving force $\mathbf{d}_i$ is given by Eq. 8.7. The last two terms in the bracket have been added in order to account for thermal diffusion (see also Chapter 12). The thermal diffusion coefficients D_i^T are additional transport properties, of which only $n - 1$ are independent since they always appear as differences as in Eq. 8.8.

Equation 8.8 expresses the driving force for diffusion, $\mathbf{d}_i$, in terms of the diffusion fluxes $(\mathbf{v}_j - \mathbf{v}_i)$. Occasionally, it is desirable to express the fluxes in terms of the driving forces

$$\mathbf{j}_i = \frac{c}{\rho RT}\sum_{j=1}^{n} M_i M_j D_{ij}\mathbf{d}_j - D_i^T \nabla \ln T. \tag{8.9}$$

We should realize that Eqs. 8.8 and 8.9 represent sets of linear relations in the fluxes and the driving forces. Thus, Eq. 8.8 can be inverted or turned inside out to yield Eq. 8.9, and the D_{ij} are related in a complicated manner to the $\mathcal{D}_{ij}$. The $\mathcal{D}_{ij}$ might be expected to have a simpler physical interpretation than the D_{ij} since they are related to the pairwise friction coefficients K_{ij} by Eq. 8.5. Furthermore, the $\mathcal{D}_{ij}$ are more nearly independent of composition.

Multicomponent diffusion equations become somewhat complicated. Not all this complication is essential, that is, necessary for the statement of the physical laws. For example, Eqs. 8.8 and 8.9 each state the diffusion law, and only one is essential. Nonessential complications are introduced by the possibility of using molar or mass fluxes and of using for the reference velocity the molar average velocity, the mass average velocity, the volume average velocity, or the solvent velocity (that is, the velocity of one of the species).

Problems

8.1 For a binary system, compare the expression for multicomponent diffusion with Fick's first law.

8.2 Bird, Stewart, and Lightfoot (p. 567) [1] express the multicomponent diffusion equation as

$$\mathbf{j}_i = \mathbf{j}_i^{(x)} + \mathbf{j}_i^{(p)} + \mathbf{j}_i^{(g)} + \mathbf{j}_i^{(T)} \tag{8.10}$$

where

$$\mathbf{j}_i^{(x)} = \frac{c^2}{\rho RT}\sum_{j=1}^{n} M_i M_j D_{ij}\left[x_j \sum_{\substack{k=1\\k\neq j}}^{n} \left(\frac{\partial \bar{G}_j}{\partial x_k}\right)_{\substack{T,p,x_s\\s\neq j,k}} \nabla x_k \right]. \tag{8.11}$$

$$\mathbf{j}_i^{(p)} = \frac{c^2}{\rho RT}\sum_{j=1}^{n} M_i M_j D_{ij}\left[x_j M_j \left(\frac{\bar{V}_j}{M_j} - \frac{1}{\rho}\right)\nabla p \right]. \tag{8.12}$$

$$\mathbf{j}_i^{(g)} = -\frac{c^2}{\rho RT}\sum_{j=1}^{n} M_i M_j D_{ij}\left[x_j M_j \left(\mathbf{g}_j - \sum_{k=1}^{n} \frac{\rho_k}{\rho}\mathbf{g}_k\right)\right]. \tag{8.13}$$

$$\mathbf{j}_i^{(T)} = -D_i^T \nabla \ln T. \tag{8.14}$$

Since there are not n^2 independent diffusion coefficients D_{ij}, the following conditions apply

$$D_{ij} = 0 \tag{8.15}$$

$$\sum_{i=1}^{n} \{ M_i M_h D_{ih} - M_i M_k D_{ik} \} = 0. \tag{8.16}$$

There is also an implied relationship among the thermal diffusion coefficients:

$$\sum_i D_i^T = 0.$$

(a) In the present text, the view is usually taken that the external force on all species is the same (see Ref. [2]):

$$\mathbf{g}_i = \mathbf{g}.$$

Show that Eqs. 8.10 through 8.14 are equivalent to Eq. 8.9.

(b) Show that the mass fluxes relative to the mass average velocity given by Eqs. 8.10 through 8.14 sum to zero.

8.3 For an ideal gas mixture, obtain the Stefan–Maxwell equation from the multicomponent diffusion equation 8.7.

8.4 How can the multicomponent diffusion equation 8.7 be simplified for an extremely dilute solution of several minor components in a solvent? Let the solvent be denoted by the subscript zero, and show that the concentration of any minor component obeys the equation

$$\frac{\partial c_i}{\partial t} + \mathbf{v} \cdot \nabla c_i = \mathcal{D}_{io} \nabla^2 c_i .$$

8.5 (a) The inverted equation of multicomponent diffusion can be written, in the absence of thermal diffusion,

$$\mathbf{j}_i = -\frac{c}{\rho RT} \sum_{j=1}^{n} M_i M_j D_{ij} \mathbf{d}_j ,$$

where the diffusion coefficients D_{ij} are subject to the restrictive conditions

$$D_{ii} = 0$$

and

$$\sum_{i=1}^{n} (M_i M_h D_{ih} - M_i M_k D_{ik}) = 0.$$

How many independent diffusion coefficients D_{ij} does this imply? Indicate your reasoning and rationalize your answer.

(b) For a three-component mixture, show that

$$D_{12} = \mathcal{D}_{12} \left[1 + x_3 \frac{\mathcal{D}_{13} M_3 / M_2 - \mathcal{D}_{12}}{x_1 \mathcal{D}_{23} + x_2 \mathcal{D}_{13} + x_3 \mathcal{D}_{12}} \right],$$

where the $\mathcal{D}_{ij}$ are the multicomponent diffusion coefficients appearing in the equation

$$\mathbf{d}_i = RT \sum_{j=1}^{n} \frac{c_i c_j}{c \mathcal{D}_{ij}} (\mathbf{v}_j - \mathbf{v}_i), \quad \text{with } \mathcal{D}_{ij} = \mathcal{D}_{ji} .$$

Is the derivation restricted to ideal-gas mixtures? How many independent D_{ij} and $\mathcal{D}_{ij}$ exist for this three-component mixture?

8.6 The multicomponent diffusion equation

$$c_i \nabla \mu_i = RT \sum_j \frac{c_i c_j}{c\mathcal{D}_{ij}} (\mathbf{v}_j - \mathbf{v}_i) \tag{8.17}$$

is macroscopic and does not depend on precisely what molecular species are present. Consider possible treatments of the binary mixture of A and B, where B can dissociate into two molecules of C:

$$B \rightleftarrows 2C. \tag{8.18}$$

In all cases, the dissociation equilibrium is reached rapidly, or instantaneously.

From one point of view, the system is regarded as a binary mixture of A and B with a diffusion coefficient $\mathcal{D}_{AB}$ (defined by Eq. 8.17), which depends on composition. From a second point of view, the system is regarded as a ternary mixture of A, B, and C but subject to the equilibrium restriction of Eq. 8.18. If this is the correct molecular picture, the three diffusion coefficients $\mathcal{D}^*_{AB}$, $\mathcal{D}^*_{AC}$, and $\mathcal{D}^*_{BC}$ (the asterisk denotes the ternary viewpoint) might be expected to be less dependent on composition than $\mathcal{D}_{AB}$. The objective is to relate $\mathcal{D}_{AB}$ to $\mathcal{D}^*_{AB}$, $\mathcal{D}^*_{AC}$, and $\mathcal{D}^*_{BC}$ in order to get some insight into the composition dependence of $\mathcal{D}_{AB}$.

(a) Relate the stoichiometric or analytical concentration c_A and c_B to the concentrations c^*_A, c^*_B, and c^*_C of the ternary viewpoint. Note that $c = c_A + c_B$ while $c^* = c^*_A + c^*_B + c^*_C$. Relate the overall fluxes $\mathbf{N}_A = c_A\mathbf{v}_A$ and $\mathbf{N}_B = c_B\mathbf{v}_B$ to the fluxes $\mathbf{N}^*_A = c^*_A\mathbf{v}_A$, $\mathbf{N}^*_B = c^*_B\mathbf{v}_B$, and $\mathbf{N}^*_C = c^*_C\mathbf{v}^*_C$ of the ternary viewpoint. In a similar manner, relate the species velocities in the two systems.

(b) The equilibrium 8.18 implies what relation between the chemical potentials μ^*_B and μ^*_C? The relationships between the overall chemical potentials and those from the ternary viewpoint are

$$c_A\mu_A = c^*_A\mu^*_A \text{ and } c_B\mu_B = c^*_B\mu^*_B + c^*_C\mu^*_C. \tag{8.19}$$

Show that $\mu_B = \mu^*_B$. Show that if the Gibbs–Duhem relation applies to the ternary viewpoint,

$$c^*_A\nabla\mu^*_A + c^*_B\nabla\mu^*_B + c^*_C\nabla\mu^*_C = 0, \tag{8.20}$$

then it also applies to the binary viewpoint,

$$c_A \nabla \mu_A + c_B \nabla \mu_B = 0. \tag{8.21}$$

(c) Apply Eq. 8.17 to the ternary system and obtain a direct relationship between the diffusion velocities $(\mathbf{v}_B^* - \mathbf{v}_A^*)$ and $(\mathbf{v}_C^* - \mathbf{v}_A^*)$.

(d) Indicate how you would now go about relating $\mathcal{D}_{AB}$ to $\mathcal{D}_{AB}^*$, $\mathcal{D}_{AC}^*$, and $\mathcal{D}_{BC}^*$. You need not carry out the algebra, but the result is

$$\mathcal{D}_{AB} = \frac{c^*}{c_B c} \frac{c_A c_B^* \mathcal{D}_{AB}^* \mathcal{D}_{BC}^* + \frac{1}{4} c_A c_C^* \mathcal{D}_{AC}^* \mathcal{D}_{BC}^* + c_B^{*2} \mathcal{D}_{AB}^* \mathcal{D}_{AC}^*}{c_A^* \mathcal{D}_{BC}^* + c_B^* \mathcal{D}_{AC}^* + c_C^* \mathcal{D}_{AB}^*} \tag{8.22}$$

(e) In order for Eq. 8.22 to give an explicit dependence of $\mathcal{D}_{AB}$ on c_A and c_B, it is necessary to have an additional relation among the concentrations. For an ideal mixture, one might define a dissociation constant

$$K = c_C^{*2} / c_B^* \tag{8.23}$$

according to the law of mass action. Then, with four parameters, $\mathcal{D}_{AB}^*$, $\mathcal{D}_{AC}^*$, $\mathcal{D}_{BC}^*$, and K, it should be possible to fit a variety of concentration dependences. How could we obtain an independent determination of K by measuring directly c_C^* and c_B^*?

8.7 (a) Show that, for gravitational equilibrium in a region of uniform temperature, the variation of the chemical potential of a species is given by

$$\nabla \mu_i = \frac{M_i}{\rho} \nabla p.$$

Do this by consideration of a reversible process of removing a mole of the species at one point in the field, moving it against the force of gravity, and reintroducing it at another point. If possible, avoid assuming that the gravitational field is uniform.

Consideration of this process gives some justification for the expression for the driving force for diffusion,

$$\mathbf{d}_i = c_i \left(\nabla \mu_i + \bar{S}_i \nabla T - \frac{M_i}{\rho} \nabla p \right),$$

since this driving force should reduce to zero in such an equilibrium situation.

(b) Consider whether this relation correctly describes the equilibrium distributions of concentration in a centrifuge.

(c) The molar mass of NaCl is 58.44 and that of H_2O is 18.015. If seawater is 0.5 $\underline{M}$ in NaCl and 55 $\underline{M}$ in H_2O at the surface, would the equilibrium concentrations of both H_2O and NaCl be higher at a depth of 1 mile?

8.8 A gas mixture has three components, which we shall designate *A*, *B*, and *C*. It is desired to react component *A* at a solid surface. However, component *B*, which is present in a small amount, also reacts rapidly at the surface. Estimate how much *B* will contaminate the deposit as a function of the rate of deposition of *A*. Component *C* does not react. As a model, assume there is an unstirred layer of thickness δ through which *A* and *B* diffuse to the surface.

(a) Demonstrate that it is plausible to assume that the fluxes are constant within this layer. Which fluxes are known in advance, and which are unknown?

(b) Set up multicomponent diffusion equations that are suitable for the determination of the composition distribution within the diffusion layer.

(c) Specify enough boundary conditions to permit the solution to these differential equations and also any undetermined constants that might appear in the answer to Part (b).

(d) Solve this problem for the concentration profiles. To simplify matters, you can now assume that all the binary diffusion coefficients $\mathcal{D}_{ij}$ take on the same numerical value.

(e) What is the maximum permissible rate of deposition of *A*? Under that condition, what is the extent of contamination of the deposit with component *B*?

References

1. R. Byron Bird, Warren E. Stewart, and Edwin N. Lightfoot. *Transport Phenomena*. New York: John Wiley & Sons, 1960.

2. John Newman and Nitash P. Balsara. *Electrochemical Systems*, 4th Edition. Hoboken, New Jersey: John Wiley and Sons, 2020.

Chapter 9

Heat Transfer in Mixtures

The thermal energy equation 4.2 can be converted by means of thermodynamic relationships into an equation for the temperature, just as we have already done for a pure fluid (see Eq. 4.3). The principal thermodynamic relationship can be expressed as

$$dU=\left(\frac{\partial U}{\partial V}\right)_{T,n_i}dV+\left(\frac{\partial U}{\partial T}\right)_{V,n_i}dT+\sum_i\left(\frac{\partial U}{\partial n_i}\right)_{\substack{T,V,n_j\\ j\neq i}}dn_i$$

$$=\left(-p+T\left(\frac{\partial p}{\partial T}\right)_{V,n_i}\right)dV+C_VdT+\sum_i\left[\overline{U}_i+\overline{V}_i\left(p-T\left(\frac{\partial p}{\partial T}\right)_{V,n_j}\right)\right]dn_i,$$

and the resulting equation for the temperature is

$$\rho\hat{C}_V\frac{DT}{Dt}=-\nabla\cdot\mathbf{q}-\boldsymbol{\tau}:\nabla\mathbf{v}-T\left(\frac{\partial p}{\partial T}\right)_{\rho_i}\nabla\cdot\mathbf{v}$$
$$+\sum_i\left\{\overline{U}_i+\overline{V}_i\left[p-T\left(\frac{\partial p}{\partial T}\right)_{\rho_j}\right]\right\}(\nabla\cdot\mathbf{J}_i-R_i). \tag{9.1}$$

The last term is zero in a pure fluid.

The heat flux can be expressed as

$$\mathbf{q}=\sum_i\overline{H}_i\mathbf{J}_i-k\nabla T+\mathbf{q}^{(x)}. \tag{9.2}$$

The Newman Lectures on Transport Phenomena
John Newman and Vincent Battaglia

ISBN 978-981-4774-27-7 (Hardcover), 978-1-315-10829-2 (eBook)
www.jennystanford.com

This might be called the heat flux with respect to the mass average velocity. The first term represents heat carried by the interdiffusion of the species. The second term represents heat transfer by conduction. The last term is the Dufour energy flux and takes care of the balance of the heat induced by the interdiffusion of the species. This becomes

$$\mathbf{q}^{(x)} = -\sum_{i=1}^{n} \frac{D_i^T}{\rho_i} \mathbf{d}_i. \tag{9.3}$$

The justification of these expressions for $\mathbf{q}$ is not readily apparent. The justification really comes from the study of the thermodynamics of irreversible processes. Note that the same thermal diffusion coefficients enter into the Dufour energy flux as in the multicomponent diffusion equation 8.8.

The term $\sum_i \overline{H}_i \mathbf{J}_i$ can be quite important. Even though we consider the external force per unit mass $\mathbf{g}$ to be the same for each species, the terms necessary for an electrolytic system are not being ignored. For example, for steady conduction in a medium of uniform composition (including the piece of copper of Chapter 7), we have

$$\nabla \cdot \sum_i \overline{H}_i \mathbf{J}_i = \mathbf{i} \cdot \nabla \Phi. \tag{9.4}$$

This is where the Joule heating enters into Eq. 9.1.

Chapter 10

Transport Properties

In the preceding chapters, we have investigated how the fluxes are related to the driving forces, and in the process, we have generated or defined certain transport coefficients that are state functions. Table 10.1 summarizes these relations and indicates the number of transport coefficients so defined. Note that $\mathcal{D}_{jk} = \mathcal{D}_{kj}$ and that $\mathcal{D}_{kk}$ is not defined. Thus, there is one $\mathcal{D}_{jk}$ for each pair of species:

n	number of $\mathcal{D}_{jk}$
1	0
2	1
3	3
4	6
5	10

Only differences in the D_j^T are defined, so that there are $n - 1$ of these coefficients.

Transport phenomena is a science somewhat akin to thermodynamics. Just as thermodynamics defines certain thermodynamic properties that are functions of ρ_i and T, so transport phenomenon defines certain transport properties that are functions of ρ_i and T. Just as we should expect a thermodynamicist to tell us

The Newman Lectures on Transport Phenomena
John Newman and Vincent Battaglia

ISBN 978-981-4774-27-7 (Hardcover), 978-1-315-10829-2 (eBook)
www.jennystanford.com

whether he/she measured C_V or C_p, we expect an experimentalist in mass transfer to tell us just which diffusion coefficient he/she measured. Unfortunately, this is not always the case.

Table 10.1 Relation of fluxes and driving forces.

Flux equation	Transport coefficients	Number†
$\boldsymbol{\tau}=-\mu\left[\nabla\mathbf{v}+(\nabla\mathbf{v})^*\right]+\left(\frac{2}{3}\mu-\kappa\right)\mathbf{I}\nabla\cdot\mathbf{v}$	$\mu=\mu(\rho_i, T)$	1
	$\kappa=\kappa(\rho_i, T)$	1
$\mathbf{q}=-k\nabla T+\sum_i\left(\overline{H}_i\mathbf{J}_i-\frac{D_i^T}{\rho_i}\mathbf{d}_i\right)$	$k=k(\rho_i, T)$	1
	$D_j^T=D_j^T(\rho_i, T)$	$n-1$
$\mathbf{d}_i=RT\sum_j\frac{c_ic_j}{c\mathcal{D}_{ij}}\left[\mathbf{v}_j-\mathbf{v}_i+\left(\frac{D_j^T}{\rho_j}-\frac{D_i^T}{\rho_i}\right)\nabla\ln T\right]$	$\mathcal{D}_{jk}=\mathcal{D}_{jk}(\rho_i, T)$	$\frac{1}{2}n(n-1)$

†where n is the number of species present.

Similarly, we want to make a distinction between the macroscopic definition of transport or thermodynamic properties and the microscopic, theoretical explanation of the variation of these properties with temperature, pressure, and composition. Both thermodynamics and transport phenomena are macroscopic sciences, which will survive the vicissitudes or changing fortunes of the microscopic theories. (Thinking of nonnewtonian fluids, we should hasten to point out that the linear flux–force laws frequently used in transport phenomena are not exact in the same sense as the laws of thermodynamics.)

The behavior of the transport properties is summarized sketchily in Table 10.2.

First consider the transport properties of dilute gases. Here the microscopic theory is known as the kinetic theory of gases and is quite successful. In dilute gases, the molecules are far apart, that is, the mean free path is long compared to the size of a molecule. (A molecule is on the order of 10^{-8} cm. The average distance between molecules at normal temperatures and pressures is about 3×10^{-7} cm, while the mean free path is on the order of 10^{-5} cm.) A study of the equilibrium state reveals that the Maxwell–Boltzmann velocity distribution applies

$$f(\mathbf{v}) = Ae^{-mv^2/2\kappa T}. \tag{10.1}$$

Table 10.2 Behavior of transport properties.

A. In dilute gases and gas mixtures		
$\mu = \dfrac{\text{const.}\sqrt{MT}}{\sigma^2\Omega}$	$k = \dfrac{\text{const.}\sqrt{T/M}}{\sigma^2\Omega}$ (monatomic)	$\mathcal{D}_{AB} = \dfrac{\text{const.}\sqrt{T^3\left(\dfrac{1}{M_A}+\dfrac{1}{M_B}\right)}}{p\sigma^2{}_{AB}\Omega_{\mathcal{D}}}$
	$k = \mu\left(\hat{C}_p + \dfrac{5}{4}\dfrac{R}{M}\right)$ (polyatomic)	
The Ω's are functions of, say, $\kappa T/\epsilon$.		
B. In liquids		
$\mu = \rho A e^{E/RT}$	$k = 2.80(cL)^{2/3}k\upsilon_s$	$\dfrac{\mu\mathcal{D}_{AB}}{\kappa T} = \text{const.} \approx \dfrac{1}{R_A}$
υ_s is the velocity of sound.		

The molecules are moving rapidly and randomly with an average velocity given by

$$\bar{u} = \frac{\int v f(\mathbf{v}) d^3 v}{\int f(\mathbf{v}) d^3 v} = \sqrt{\frac{8\kappa T}{\pi m}}. \tag{10.2}$$

The study of transport processes becomes more complicated than the study of the equilibrium state. For a gas whose molecules are hard spheres of diameter d, the mean free path becomes

$$\lambda = \frac{1}{\sqrt{2}\,\pi d^2 n}. \tag{10.3}$$

The transport properties of hard sphere molecules are

$$\nu = \frac{\mu}{\rho} = \frac{k}{\rho\hat{C}_V} = \mathcal{D}_{AA^*} = \frac{1}{3}\bar{u}\lambda = \frac{2}{3\pi n d^2}\sqrt{\frac{\kappa T}{\pi m}}. \tag{10.4}$$

Here ν is the kinematic viscosity, k is the thermal conductivity, and $\mathcal{D}_{AA^*}$ is the self-diffusion coefficient for the interdiffusion of two hard sphere species A and A^* of the same diameter and mass. Since $\rho = Mp/RT$, the quantities in Eq. 10.4 are proportional to $T^{3/2}$ and inversely proportional to the pressure, while μ and k are independent of p and proportional to $T^{1/2}$.

The hard sphere model is not a very good representation of our idea of intermolecular forces (see Fig. 10.1). Therefore, various semiempirical force laws have been introduced, such as the Lennard–Jones "6-12" potential:

$$\phi(r) = 4\epsilon\left[\left(\frac{\sigma}{r}\right)^{12} - \left(\frac{\sigma}{r}\right)^{6}\right]. \tag{10.5}$$

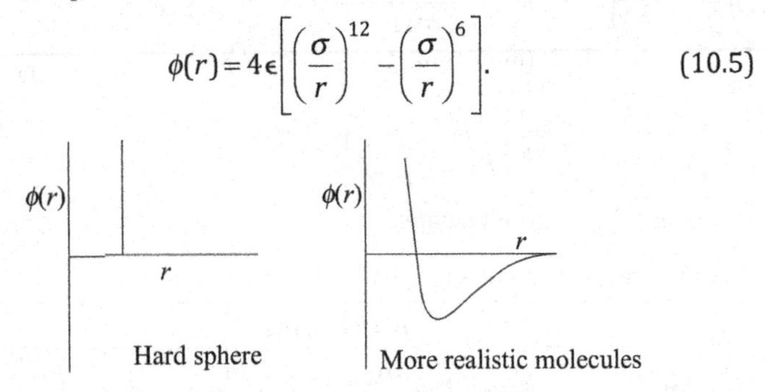

Figure 10.1 Intermolecular forces can be represented by the potential energy of interaction $\phi(r) = \int_r^\infty f(r)\cdot dr$ as long as the molecules have spherical symmetry.

When these are used in the kinetic theory of gases, one then obtains, for example,

$$\mu \approx c\mathcal{D}_{AA^*} \approx \text{const.}\frac{\sqrt{T/M}}{\sigma^2\Omega(\kappa T/\epsilon)}. \tag{10.6}$$

Because of the temperature dependence of the Ω, these increase more rapidly with T than $T^{1/2}$ and are in better agreement with experimental data. Ref. [1] gives a detailed account of the theory of transport in dilute gases.

For liquids, the viscosity is not very sensitive to pressure, and it decreases with temperature like

$$\nu = \text{const. } e^{\Delta G_o^{\ddagger}/RT}. \tag{10.7}$$

For diffusion, the Stokes–Einstein relation is useful

$$\frac{\mathcal{D}_{AB}\mu}{\kappa T} = \text{const.} \approx \frac{1}{R_A}. \tag{10.8}$$

Here R_A is a length on the order of the radius of a molecule. The thermal conductivity does not change nearly so rapidly,

$$k = 2.80\left(\frac{c}{L}\right)^{3/2} \kappa v_s \tag{10.9}$$

where v_s is the velocity of sound. Thus, both μ and $\mathcal{D}_{AB}$ show a strong dependence on temperature, while k does not and may in fact increase or decrease with temperature.

For dilute gases, the viscosity increases with temperature, but for liquids, it decreases. This is due to the different types of molecular motion and the different mechanisms of momentum transfer. In gases, there is free flight, while in liquids, there is bumping. This also shows up in the diffusion coefficient. For gases,

$$c\mathcal{D}_{AB} \propto \mu,$$

but for liquids

$$\mathcal{D}_{AB} \propto \frac{1}{\mu}.$$

In dilute gases, free flight is the mechanism of both mass and momentum transport, and bumping terminates the process. Hence, $c\mathcal{D}_{AB} \propto \mu$. For liquids, bumping promotes momentum transport but not mass transport. Bumping is a drag on the particles, and $\mathcal{D}_{AB} \propto 1/\mu$. Ref. [2] covers transport and thermodynamic properties in general.

The subject of transport properties is an important aspect of transport phenomena and is worthy of more than passing attention.

(a) Use. If you need to know a transport property you may be able to look up the measured value or you can probably make a good estimate from correlations.

(b) Measurements, of course, provide the basic data. Hence the development of experimental methods is important. If possible, clearly define the properties for which you report values.

(c) Empirical correlations are useful to
- condense data.
- allow extrapolation to unmeasured conditions.
- give an idea of trends for theoretical interpretation.

(d) Theoretical interpretation. The microscopic theories are important not only to predict transport properties and guide their correlation but also to demonstrate our knowledge of molecular mechanics. We can also get information about molecular structure and molecular interactions from transport properties. For example, Lennard–Jones parameters can be obtained from viscosity data as well as from second virial coefficients. There should be some union of thermodynamic and transport properties in their microscopic explanation.

Problem

10.1 The viscosity for liquids is not very sensitive to pressure. Do you expect the viscosity for gases to be sensitive to pressure change? Why? Justify your answer in terms of mechanisms of viscous momentum transfer.

References

1. Sydney Chapman and T. G. Cowling. *The Mathematical Theory of Non-Uniform Gases.* Cambridge: The University Press, 1970.
2. Bruce E. Poling, John M. Prausnitz, and John. P. O'Connell, *The Properties of Gases and Liquids.* New York: McGraw-Hill Book Company, 2001.

Chapter 11

Entropy Production

The thermal energy equation 4.2 can be converted by means of thermodynamic relationships into an equation for the entropy. The variation of internal energy is

$$dU = TdS - pdV + \sum_i \mu_i dn_i$$

or

$$\frac{D\hat{U}}{Dt} = T\frac{D\hat{S}}{Dt} - p\frac{D\hat{V}}{Dt} + \sum_i \mu_i \frac{D\hat{n}_i}{Dt}.$$

Substitution into Eq. 4.2 yields

$$\rho T\frac{D\hat{S}}{Dt} = -\nabla\cdot\mathbf{q} - \boldsymbol{\tau}:\nabla\mathbf{v} + \sum_i \mu_i[\nabla\cdot\mathbf{J}_i - R_i]. \tag{11.1}$$

By using the Gibbs–Duhem relation and the relation

$$\overline{H}_i - \mu_i = T\overline{S}_i,$$

Eq. 11.1 can be rearranged to read

$$\boxed{\begin{aligned}&\frac{\partial\rho\hat{S}}{\partial t} + \nabla\cdot\left[\frac{1}{T}\mathbf{q}' + \sum_i \overline{S}_i\mathbf{N}_i\right] = g\\ &= \frac{1}{T}\left\{-\boldsymbol{\tau}:\nabla\mathbf{v} - \sum_i \mu_i R_i - \mathbf{q}'\cdot\nabla\ln T - \sum_i \mathbf{v}_i\cdot\mathbf{d}_i\right\},\end{aligned}} \tag{11.2}$$

The Newman Lectures on Transport Phenomena
John Newman and Vincent Battaglia

ISBN 978-981-4774-27-7 (Hardcover), 978-1-315-10829-2 (eBook)
www.jennystanford.com

where

$$\mathbf{q}' = \mathbf{q} - \sum_i \overline{H}_i \mathbf{J}_i = -k\nabla T + \mathbf{q}^{(x)}, \tag{11.3}$$

and where $\mathbf{d}_i$ is given by Eq. 8.7.

The first term in the entropy Eq. 11.2 is obviously the accumulation of entropy per unit volume. The second term can be regarded as the divergence of the entropy flux. The term on the right, g, is the rate of production of entropy per unit volume by irreversible processes. These are as follows:

term	source of entropy production
$-\frac{1}{T}\boldsymbol{\tau}:\nabla\mathbf{v}$	viscous dissipation
$-\frac{1}{T}\sum_i \mu_i R_i$	chemical reactions occurring irreversibly
$-\frac{1}{T}\mathbf{q}'\cdot\nabla \ln T$	heat conduction
$-\frac{1}{T}\sum_i \mathbf{v}_i\cdot\mathbf{d}_i$	diffusion

A statement of the second law of thermodynamics is that any irreversible process must produce entropy, or, in our case, $g \geq 0$. Let us examine some consequences of the second law of thermodynamics.

For a pure fluid, the entropy production is

$$g = -\frac{1}{T}\boldsymbol{\tau}:\nabla\mathbf{v} - \frac{1}{T}\mathbf{q}'\cdot\nabla \ln T.$$

The introduction of Newton's law of viscosity 2.3 and Fourier's law of heat conduction 4.4 gives

$$g = \frac{\mu}{T}\Phi_V + \frac{k}{T^2}(\nabla T)^2,$$

where $$\Phi_V = \left[\nabla\mathbf{v} + (\nabla\mathbf{v})^*\right]:\nabla\mathbf{v} - \frac{2}{3}(\nabla\cdot\mathbf{v})^2.$$

Since $g \geq 0$ and since Φ_V and $(\nabla T)^2$ are nonnegative, the viscosity and thermal conductivity must also be nonnegative:

$$\mu \geq 0 \quad \text{and} \quad k \geq 0.$$

One could also show that the bulk viscosity k is nonnegative.

For a binary system with no chemical reactions, the entropy production is given by

$$Tg = -\boldsymbol{\tau} : \nabla\mathbf{v} - \mathbf{q}' \cdot \nabla \ln T - (\mathbf{v}_A - \mathbf{v}_B) \cdot \mathbf{d}_A.$$

For this system, the diffusion law 8.8 is

$$\mathbf{v}_A - \mathbf{v}_B = -\frac{c\mathcal{D}_{AB}}{c_A c_B RT}\mathbf{d}_A - \left(\frac{D_A^T}{\rho_A} - \frac{D_B^T}{\rho_B}\right)\nabla \ln T,$$

and the heat-conduction law is

$$\mathbf{q}' = -\left(\frac{D_A^T}{\rho_A} - \frac{D_B^T}{\rho_B}\right)\mathbf{d}_A - kT\nabla \ln T.$$

The fact that $g \geq 0$ implies that

$$\mu \geq 0, \quad \kappa \geq 0, \quad k \geq 0, \quad \mathcal{D}_{AB} \geq 0,$$

and

$$\left(\frac{D_A^T}{\rho_A} - \frac{D_B^T}{\rho_B}\right)^2 \leq \frac{kc\mathcal{D}_{AB}}{c_A c_B R}.$$

For a ternary system at constant temperature and pressure,

$$gT = \frac{RT}{c}\left[\frac{c_1 c_2}{\mathcal{D}_{12}}(\mathbf{v}_2 - \mathbf{v}_1)^2 + \frac{c_1 c_3}{\mathcal{D}_{13}}(\mathbf{v}_3 - \mathbf{v}_1)^2 + \frac{c_2 c_3}{\mathcal{D}_{23}}(\mathbf{v}_3 - \mathbf{v}_2)^2\right],$$

where viscous dissipation has been neglected. The implications are left to the exercises.

Problems

11.1 For a hypothetical, three-component liquid phase, the following diffusion coefficients are reported:

$$\left.\begin{aligned} \mathcal{D}_{12} &= 0.2273 \times 10^{-5}\ \text{cm}^2/\text{s} \\ \mathcal{D}_{13} &= 5.00 \times 10^{-5}\ \text{cm}^2/\text{s} \\ \mathcal{D}_{23} &= 0.4545 \times 10^{-5}\ \text{cm}^2/\text{s} \end{aligned}\right\} \quad \begin{aligned} &\text{for } x_1 = x_2 = 0.0833, \\ &x_3 = 0.833, \\ &T = 25°\text{C}, p = 1\ \text{atm}. \end{aligned}$$

Are these values consistent with the second law of thermodynamics? Justify your answer.

The diffusion coefficients reported above are those which appear in the expression

$$c_i \nabla \mu_i = RT \sum_j \frac{c_i c_j}{c\mathcal{D}_{ij}} (\mathbf{v}_j - \mathbf{v}_i).$$

11.2 Show that for a binary mixture in the absence of chemical reactions, the second law of thermodynamics implies that $\mu \geq 0$, k (thermal conductivity) ≥ 0, $\mathcal{D}_{AB} \geq 0$, and

$$\left(\frac{D_A^T}{\rho_A} - \frac{D_B^T}{\rho_B} \right) \leq \frac{kc\mathcal{D}_{AB}}{c_A c_B R}.$$

Here the diffusion law is expressed as

$$\mathbf{v}_A - \mathbf{v}_B = -\frac{c\mathcal{D}_{AB}}{c_A c_B RT} \mathbf{d}_A - \left(\frac{D_A^T}{\rho_A} - \frac{D_B^T}{\rho_B} \right) \nabla \ln T,$$

and the heat flux is

$$\mathbf{q} = \overline{H}_A \mathbf{J}_A + \overline{H}_B \mathbf{J}_B - k\nabla T - \left(\frac{D_A^T}{\rho_A} - \frac{D_B^T}{\rho_B} \right) \mathbf{d}_A.$$

11.3 For a binary liquid system, it is customary to define the Soret coefficient σ as

$$\sigma = \frac{1}{DT} \left(\frac{D_A^T}{\rho_A} - \frac{D_B^T}{\rho_B} \right),$$

where B is the solute, A is the solvent, and D is the measured diffusion coefficient. (D is the same as $\mathcal{D}_{AB}$ in Chapter 6. You can take the solution to be ideal, in which case D is also the same as $\mathcal{D}_{AB}$ in Chapter 8.) Our experimental system consists of an aqueous solution of potassium chloride (component B) confined between two horizontal inert plates, the upper being maintained at the hot temperature T_H and the lower at the cold temperature T_C. Take σ to be -0.003 (K)$^{-1}$, D to be 2×10^{-5} cm^2/s, and $\alpha = k / \rho \hat{C}_p = 1.43 \times 10^{-3} \text{cm}^2/\text{s}$.

(a) Are these values of the physical properties at 25°C in harmony with the requirements of the second law of thermodynamics? Take the molar heat capacity to be $\tilde{C}_p = 9R$.

(b) For this negative Soret coefficient, will thermal diffusion tend to concentrate the solute at the hot surface or the cold surface?

(c) With neglect of pressure gradients, show that the solute flux is now given by

$$\mathbf{J}_B^* = -cD\left(\nabla x_B - x_A x_B \sigma \nabla T\right).$$

(d) For the situation where a steady state has developed, use the differential material balances to show that $\mathbf{v}_A = \mathbf{v}_B = 0$.

(e) Integrate the flux equation for this steady state. If $x_B = 0.01$ adjacent to the cold plate at 0°C, what will be the value of x_B adjacent to the hot plate at 50°C?

11.4 A binary mixture of A and B is forced to flow through a reverse-osmosis membrane by means of a substantial pressure difference $p_1 - p_2$. (We might note, incidentally, that the composition of the solution is different on the two sides of the membrane. There is necessarily a third stream leaving the system. This leaves at the same pressure and temperature as stream 1. It is assumed here that this third stream is large enough and the stirring on the upstream side of the membrane is adequate so that the composition on the upstream side of the membrane is the same as that of the entering stream 1.)

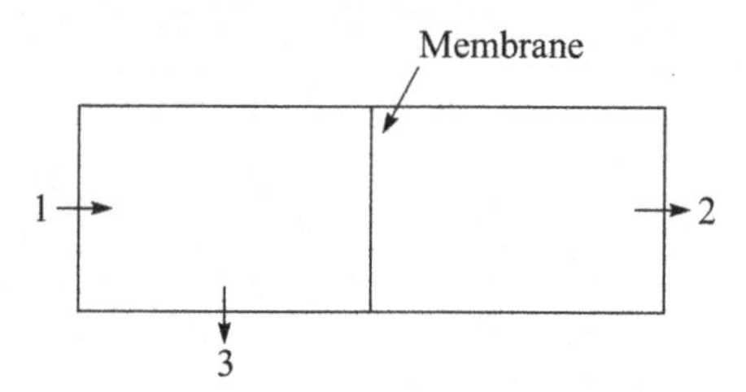

For an isothermal system of temperature T, the rate of entropy production within the membrane is

$$gL = N_A \frac{\mu_A^{(1)} - \mu_A^{(2)}}{T} + N_B \frac{\mu_B^{(1)} - \mu_B^{(2)}}{T},$$

where N_A and N_B are the fluxes through the membrane and where L is the thickness of the membrane. These fluxes are, in turn, related to the driving forces through the relations

$$N_A = L_A\left(\mu_A^{(1)} - \mu_A^{(2)}\right) + L_{AB}\left(\mu_B^{(1)} - \mu_B^{(2)}\right),$$

$$N_B = L_{BA}\left(\mu_A^{(1)} - \mu_A^{(2)}\right) + L_B\left(\mu_B^{(1)} - \mu_B^{(2)}\right).$$

No relationship between L_{AB} and L_{BA} is given.

What restrictions on the coefficients L_A, L_B, L_{AB}, and L_{BA} are placed by the second law of thermodynamics?

Chapter 12

Coupled Transport Processes

Let us discuss now the subject known as nonequilibrium thermodynamics or irreversible thermodynamics. The scope of this field is about the same as that of transport phenomena, but the manner of approach is considerably different. The choice of the titles, nonequilibrium thermodynamics and irreversible thermodynamics, is also unfortunate, particularly after one has learned that thermodynamics deals not with dynamic situations but with equilibrium or reversible processes.

Nonequilibrium thermodynamics deals first of all with the proper macroscopic variables (such as temperature, pressure, and density) and their definition in nonequilibrium situations. The same discussion must also underlie transport phenomena. Nonequilibrium thermodynamics next treats the rate laws for

- the viscous stress.
- mass transfer of the several species.
- transport of energy.
- chemical reactions.

The treatment is usually restricted to a linear dependence of the rates (loosely referred to as fluxes) on the relevant driving forces. Thus, Newton's law of viscosity, Fourier's law of heat conduction, and Fick's law of diffusion are all linear relations between fluxes

The Newman Lectures on Transport Phenomena
John Newman and Vincent Battaglia

ISBN 978-981-4774-27-7 (Hardcover), 978-1-315-10829-2 (eBook)
www.jennystanford.com

and driving forces. Stress relations for nonnewtonian fluids are then excluded. The linear laws are adequate for heat and mass transfer since nonlinear conduction and diffusion have not been documented experimentally.† The restriction to linear rate laws for chemical reactions can be quite confining since many reactions are not first order. Linear approximations to the rate laws will, however, be applicable close to equilibrium.

There is an emphasis in nonequilibrium thermodynamics on the coupling among different rate processes. This refers to the fact that a temperature gradient can lead to a mass flux or the fact that a temperature gradient in one direction in an anisotropic crystal can give rise to a heat flux in another direction. Fortunately, it is found that the coupling is limited. The rates and driving forces of several chemical reactions may be coupled among themselves, but they are not directly related to the transport of momentum, energy, and mass. Mass and energy transport, being of the same tensorial order, can be coupled, but they should be independent of both the stress and the chemical reactions.

Finally, nonequilibrium thermodynamics emphasizes symmetry relations. These are principally the *Onsager reciprocal relations.* They refer, for example, to the fact that the thermal-diffusion coefficients are the same in both Eq. 8.8 for mass transfer and Eq. 9.3 for heat transfer. The reciprocal relations for multicomponent diffusion are already included in the statement of Eq. 8.5 that $\mathcal{D}_{ij} = \mathcal{D}_{ij}$, arrived at by a heuristic argument invoking Newton's third law of motion.

Thus, although we have not treated anisotropic media in this text, the coupling and symmetry relations for various mass fluxes and concentration gradients have been introduced in Chapter 8 on multicomponent diffusion, and the coupling between heat transfer and mass transfer has been discussed in Chapters 8 and 9. The symmetry of the viscous stress is demonstrated in Problem 2.1, and any reciprocal relation must already be built into Eq. 2.3. We shall leave reciprocal relations for chemical reactions for texts on chemical kinetics.

Once the possible coupling of heat and mass transfer has been recognized, the principal contribution of nonequilibrium thermodynamics can be seen to be the Onsager reciprocal relations.

---

†An exception would be the breakdown of Ohm's law for very high electric fields.

Their role, however, is merely to reduce the number of fundamental transport properties from n^2 to $\frac{1}{2}n(n+1)$ for a system with n species. (Here we count the thermal conductivity, the diffusion coefficients, and the thermal-diffusion coefficients.) This should reduce the experimental effort required to determine the transport properties.

The rigorous derivation of reciprocal relations is complicated and has its basis in fluctuation theory [1–3], part of statistical mechanics. This method was used by Onsager [1, 2] to demonstrate symmetry principles for heat conduction in anisotropic crystals. An attempt is made in Section 12.3 to apply this method to thermal-diffusion coefficients. Generally, the derivation of reciprocal relations would have to be regarded as inadequate. There is in the literature a lot of attention devoted to the appropriate definition of fluxes and forces for which the transport coefficients would be symmetric [4]. This usually involves examination of the form of the entropy production, as given in Eq. 11.2. It has been clearly shown by Coleman and Truesdell [5], however, that a commonly used recipe for determining the appropriate forces and fluxes has an incorrect basis. This is developed in Section 12.1.

We have attempted here to define the scope of nonequilibrium thermodynamics and to set it in perspective in relationship to transport phenomena and statistical mechanics. The whole subject can frequently be ignored in engineering work because the coupling may be weak, because there may be other ways to say the same thing, or because transport properties in multicomponent systems may be too poorly known to justify an elaborate or rigorous treatment. (In this regard, Wei [6] has expressed the viewpoint of limited practical applicability.) Nevertheless, there has been a strong and continuing interest in coupled irreversible phenomena on the part of physical and biological scientists since the 19th century.

A course of instruction for engineers should involve a practical design problem. For coupled transport processes, we suggest the design of a thermoelectric heat engine or refrigerator and the design of a thermal-diffusion column or Clausius–Dickel column for the separation of a binary mixture [7]. Thermoelectric effects are elaborated upon in Section 12.2 with this design problem as a goal.

Nonequilibrium thermodynamics is sometimes applied directly to larger systems, such as membranes, without passing through the form of differential rate laws. Imagine a porous ceramic, with straight pores of radius R, separating two compartments of pure water. A pressure difference will result in the fluid flow through the porous disk, and an electric field will produce a flow of charge. If the electric double layer at the ceramic–water interface is considered, these processes are found to be coupled. The current density, averaged over the cross section of the pore, is given by

$$< i_z > = E_z \kappa_{\text{eff}} - \frac{dp}{dz} \frac{\lambda q_2}{\mu} \frac{I_2(R/\lambda)}{I_1(R/\lambda)}, \tag{12.1}$$

where E_z is the axial electric field, κ_{eff} is the effective electric conductivity of the water in the pore, q_2 is the surface charge density in the diffuse part of the double layer, I_1 and I_2 denote the modified Bessel functions of the first kind of order one and two, respectively, and λ is the Debye length for the diffuse part of the double layer. Similarly, the average fluid velocity is

$$< v_z > = E_z \frac{\lambda q_2}{\mu} \frac{I_2(R/\lambda)}{I_1(R/\lambda)} - \frac{R^2}{8\mu} \frac{dp}{dz}. \tag{12.2}$$

The fact that the coefficient of E_z in Eq. 12.2 is the same as the coefficient of $-dp/dz$ in Eq. 12.1 can be regarded as an example of the Onsager reciprocal relations, although these equations were derived (Ref. [8], pp. 205–206) without any explicit reference to symmetry.

12.1 Entropy Production

For many years, it was thought that the matrix of transport coefficients should be symmetric as long as the forces and fluxes chosen were *conjugate* so that when multiplied and summed, the entropy-production term resulted. This was disproved in a simple manner by Coleman and Truesdell in 1960 [5].

Condition 1. Suppose there exist forces X_n and fluxes J_n such that the entropy production g is given by

$$Tg = \sum_n J_n X_n. \tag{12.3}$$

One should choose independent forces and fluxes with convenient units.†

Condition 2. Suppose that these forces and fluxes are linearly related:

$$J_n = \sum_m L_{nm} X_m. \tag{12.4}$$

Condition 3. Suppose that the Onsager reciprocal relation applies:

$$L_{mn} = L_{nm}. \tag{12.5}$$

Now construct a new set of fluxes J'_n:

$$J'_n = J_n - \sum_m W_{nm} X_m, \tag{12.6}$$

where W_{nm} is any suitable anti-symmetric matrix, that is, $W_{ij} = -W_{ji}$. By suitable we mean that the components of the matrix should have appropriate dimensions. The set of fluxes J'_n and forces X_n now has the following properties:

a. They yield the entropy production, as in Condition 1:

$$Tg = \sum_n \left(J'_n + \sum_m W_{nm} X_m \right) X_n = \sum_n J'_n X_n. \tag{12.7}$$

b. They are linearly related, as in Condition 2:

$$J'_n = \sum_m (L_{nm} - W_{nm}) X_m. \tag{12.8}$$

---

†For example, J_1, J_2, and J_3 could be the three components of the heat flux $\mathbf{q}'$ while X_1, X_2, and X_3 would be the corresponding components of $-\nabla \ln T$.

For the diffusion fluxes, we could use the components of $\mathbf{v}_i - \mathbf{v}_o$, where the velocity of species o is chosen for the reference velocity. The conjugate forces would then be the corresponding components of $-\mathbf{d}_i$.

Alternatively, the diffusion fluxes could be the components of $c_i(\mathbf{v}_i - \mathbf{v}_o)$, and the conjugate forces would be the components of $-\left(\nabla \mu_i + \bar{S}_i \nabla T - \frac{M_i}{\rho} \nabla p \right)$. By excluding explicit consideration of species o, independent forces and fluxes are obtained.

For the component τ_{xy} of the viscous momentum flux, one would use the conjugate force $-\partial v_x/\partial y$.

One must ascertain the number of independent reactions before decomposing the term $-\sum_i \mu_i R_i$ into forces and "fluxes."

c. The Onsager reciprocal relation does not apply:

$$L_{nm} - W_{nm} \neq L_{mn} - W_{mn}. \tag{12.9}$$

This theorem shows clearly that there can be no general assurance that the Onsager reciprocal relation applies when a set of forces and fluxes merely satisfy Conditions 1 and 2.

A related theorem states that it is frequently possible to construct a set of forces and fluxes for which the transport matrix is symmetric. Suppose there exist forces and fluxes satisfying Conditions 1 and 2, but the measured matrix L_{nm} may not be symmetric. Decompose this matrix into a symmetric matrix L^*_{nm} and an anti-symmetric matrix W_{nm}:

$$L_{nm} = L^*_{nm} + W_{nm}, \tag{12.10}$$

and construct a new set of fluxes J'_n according to Eq. 12.6. The fluxes J'_n and forces X_n still yield the entropy production as in Condition 1 (see Eq. 12.7) and are linearly related as in Condition 2 (see Eq. 12.8). The matrix of coefficients is now L^*_{nm} and is symmetric by construction. Hence, one can say that the Onsager reciprocal relation applies.

This theorem shows that it is trivial to state that there exists a set of forces and fluxes for which the Onsager reciprocal relation applies. Each coefficient L_{nm} needed to be measured in order to construct the appropriate forces and fluxes. Monroe and Newman [4] use Onsager's original methods of microscopic reversibility and decay of fluctuations to demonstrate how to confirm the Onsager reciprocal relations for the forces and fluxes that he identified in 1945 but did not really prove the reciprosity.

12.2 Thermoelectric Effects

Our objective here is to treat the thermocouple shown in Fig. 12.1. We permit a current to flow in the system, thereby allowing the device to operate as a heat engine or a refrigerator, depending on the direction of the current. Before we analyze the whole system, we should specialize the transport laws for this application.

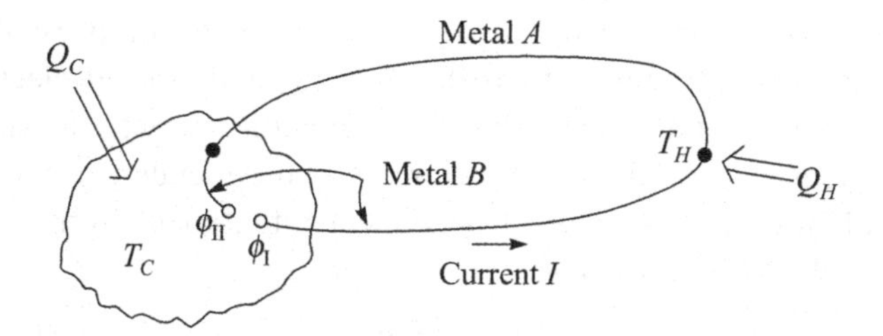

Figure 12.1 Thermoelectric heat engine, basically a thermocouple from which an electric current I is withdrawn.

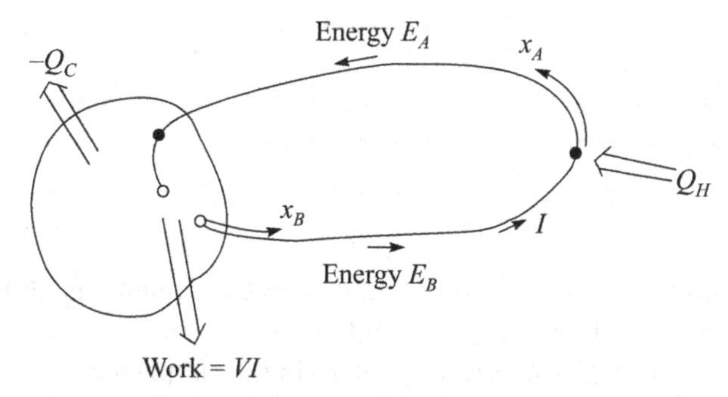

Figure 12.2 Coordinates and flow of energy in thermoelectric heat engine. E_A and E_B are the energy fluxes in the wires multiplied by the cross-sectional area of the wire.

Regard each metal to be composed of electrons (−) and positive ions (+), the latter being fixed in position. The medium is electrically neutral

$$z_+c_+ + z_-c_- = 0, \tag{12.11}$$

where z_i is the valence or charge number, and the current density can be identified as a diffusion current:

$$\mathbf{i} = F(z_+c_+\mathbf{v}_+ + z_-c_-\mathbf{v}_-) = z_+c_+F(\mathbf{v}_+ - \mathbf{v}_-), \tag{12.12}$$

where F = 96,487 C/equiv is Faraday's constant.

In the absence of a pressure gradient, the multicomponent diffusion equation 8.8 becomes

$$\mathbf{d}_- = c_-\left(\nabla\mu_- + \bar{S}_-\nabla T\right) = \frac{RTc_+c_-}{c\mathcal{D}_{+-}}\left[\mathbf{v}_+ - \mathbf{v}_- + \left(\frac{D_+^T}{\rho_+} - \frac{D_-^T}{\rho_-}\right)\nabla \ln T\right]. \tag{12.13}$$

Since electrons are a charged species, μ_- is taken to represent the *electrochemical* potential, there being a dependence on electrical state as well as on composition and temperature. For this single-component system, there is no composition dependence, and it is convenient to replace the electrochemical potential of electrons by a defined electric potential Φ:

$$\mu_- = z_- F\Phi. \tag{12.14}$$

At uniform temperature, this association with the electric potential is unambiguous, and Eq. 12.13 reduces to Ohm's law:

$$\nabla\mu_- = \frac{RT}{c\mathcal{D}_{+-}}\frac{\mathbf{i}}{z_+F} = z_-F\nabla\Phi = -\frac{z_-F}{\kappa}\mathbf{i}. \tag{12.15}$$

In this manner, we are able to identify the electric conductivity κ with the multicomponent diffusion coefficient:

$$\kappa = -z_+z_-\frac{F^2c}{RT}\mathcal{D}_{+-}. \tag{12.16}$$

For nonuniform temperature, other electric potentials different from that in Eq. 12.14 may be found in use.

Equation 12.13 can now be written in the simple form

$$\nabla\Phi = -\frac{\mathbf{i}}{\kappa} - \xi\nabla T \tag{12.17}$$

if we collect the coefficients of ∇T and define the thermoelectric coefficient

$$\xi = \frac{z_+c_+F}{\kappa T}\left(\frac{D_+^T}{\rho_+} - \frac{D_-^T}{\rho_-}\right) + \frac{\bar{S}_-}{z_-F}. \tag{12.18}$$

The separate contributions to the thermoelectric coefficient ξ would not be subject to separate experimental determination in any case.

12.2.1 Energy Transfer

Equation 4.6 expresses the total energy flux

$$\mathbf{e} = \left(\frac{1}{2}\rho v^2 + \rho\hat{U}\right)\mathbf{v} + (\boldsymbol{\tau} + p\mathbf{I})\cdot\mathbf{v} + \mathbf{q}. \tag{12.19}$$

Neglect the stress $\boldsymbol{\tau}$ and the kinetic energy and substitute Eqs. 9.2 and 9.3 to obtain

$$\mathbf{e} = \sum_i \overline{H}_i c_i \mathbf{v}_i - k\nabla T - \left(\frac{D_-^T}{\rho_-} - \frac{D_+^T}{\rho_+} \right) \mathbf{d}_- . \tag{12.20}$$

Next we want to set $\mathbf{v}_+ = 0$, introduce $\mathbf{i}$ by means of Eq. 12.12, and replace $\mathbf{d}_-$ by means of Eq. 12.13. Also use the relation

$$\overline{H}_- = \mu_- + T\overline{S}_- = z_- F\Phi + T\overline{S}_- . \tag{12.21}$$

After collection of coefficients, Eq. 12.20 then becomes

$$\mathbf{e} = (\Phi + \xi T)\mathbf{i} - k'\nabla T, \tag{12.22}$$

where

$$k' = k - \frac{(z_+ c_+ F)^2}{\kappa T} \left(\frac{D_+^T}{\rho_+} - \frac{D_-^T}{\rho_-} \right)^2 \tag{12.23}$$

is a modified thermal conductivity. In fact, k' is the value that would normally be measured and tabulated since an electric current is usually not permitted during the determination of the thermal conductivity.

We do not find it profitable to decompose the flux $\mathbf{e}$ into an electric term and a heat flux. In the presence of an electric current, we shall not even attempt to distinguish how much energy flows in wire A and how much in wire B in Fig. 12.1.

We have already utilized the Onsager reciprocal relation when we use the same thermal-diffusion coefficients D_i^T both in the multicomponent diffusion equation 12.13 and in the energy flux in Eq. 12.20. For this reason, the same thermoelectric coefficient ξ appears in the two basic practical Eqs. 12.17 and 12.22 for the analysis of thermoelectric effects.

12.2.2 Thermoelectric Equation

The potential V of the device in Fig. 12.1 is defined by

$$V = \Phi_{\mathrm{II}} - \Phi_{\mathrm{I}} = \int_{\mathrm{I}}^{\mathrm{II}} d\Phi. \tag{12.24}$$

Notice that the terminals are both at the same temperature T_C. Thus, an external device such as a voltmeter, a motor, a battery, or a resistor can be connected without any concern about the definition of the potential difference V. At the temperature T_C, a distinction is being made between the heat added, Q_C, and the work, VI, done by the system.

To evaluate Eq. 12.24, integrate Eq. 12.17 around the loop in Fig. 12.1 from I to II. The electrochemical potential of electrons is continuous across the junctions between different electronic conductors, barring any contact resistance. Consider the wires to be thin, and integrate over the cross section of a wire to obtain a resistance.

$$V = -\int_{\mathrm{I}}^{\mathrm{II}} \frac{\mathbf{i} \cdot \mathbf{dl}}{\kappa} - \int_{T_C}^{T_H} \xi_B dT - \int_{T_H}^{T_C} \xi_A dT$$

$$= \int_{T_C}^{T_H} (\xi_A - \xi_B) dT - I \oint \frac{d\ell}{\kappa A}, \tag{12.25}$$

where A is the cross-sectional area.

The first term on the right is the thermoelectric potential of the thermocouple at open circuit. This can be found tabulated in thermocouple tables. The thermoelectric coefficient can be measured only in differences like $\xi_A - \xi_B$, the difference between the values for two materials. For a similar reason, we feel that we cannot measure Φ at positions around the loop without creating another thermocouple in the measuring device.

The second term on the right in Eq. 12.25 is the ohmic potential drop, the total current times the electric resistance of the loop. The short-circuit current would be equal to the open-circuit potential divided by the resistance. The conductivities of the metals change with temperature, and the temperature distribution along the wires can change with the current level. Strictly speaking, the evaluation of the integral for the resistance is more difficult than it appears.

12.2.3 Heat Generation at an Interface

The normal components of both $\mathbf{e}$ and $\mathbf{i}$ are continuous at an interface. In most problems, Φ and T are also continuous. Let x be directed from phase A to phase B. Equation 12.22 yields

$$-k_B' \frac{\partial T_B}{\partial x} + k_A' \frac{\partial T_A}{\partial x} = T i_x (\xi_A - \xi_B). \tag{12.26}$$

This is known as the Peltier effect. Current passing across the interface gives rise to a heat generation or absorption.

12.2.4 Heat Generation in the Bulk

Both $\nabla \cdot \mathbf{e}$ and $\nabla \cdot \mathbf{i}$ are zero in the steady state. Equation 12.22 yields

$$0 = \nabla \cdot (k' \nabla T) - \mathbf{i} \cdot \nabla \Phi - \mathbf{i} \cdot \nabla (\xi T), \tag{12.27}$$

and with substitution of Eq. 12.17, we have

$$0 = \nabla \cdot (k' \nabla \mathrm{T}) + \frac{\mathbf{i} \cdot \mathbf{i}}{\kappa} - T\mathbf{i} \cdot \nabla \xi. \tag{12.28}$$

The second term on the right is the Joule effect. It always represents a dissipative effect. The last term is the Thompson effect. If ξ depends only on temperature, this term changes sign if the direction of $\mathbf{i}$ or ∇T is reversed. It is similar to the Peltier effect.

12.2.5 Thermoelectric Engine

Equation 12.22 appears to be harder to integrate, in general, than Eq. 12.17. Therefore, we shall simplify the problem by taking the physical properties κ, k', and ξ to be independent of temperature and by assuming that each wire has a uniform cross section and is thermally and electrically insulated. Gross energy balances give (see Fig. 12.2)

$$E_A = E_B + Q_H = E_B - Q_C + VI. \tag{12.29}$$

Since I is uniform along a wire, integration of Eq. 12.17 gives

$$\Phi = \frac{-I}{\kappa_B A_B} x_B - \xi_B (T - T_C) \tag{12.30}$$

in metal B, with Φ_l taken to be zero, and

$$\Phi = \frac{-I}{\kappa_B A_B} \ell_B - \frac{I}{\kappa_A A_A} x_A - \xi_B (T_H - T_C) - \xi_A (T - T_H) \tag{12.31}$$

in metal A. Substitution into Eq. 12.22 and integration gives, since E is uniform along a wire,

$$\frac{x_B}{A_B} E_B = \frac{I}{A_B} \left[\frac{-I}{\kappa_B A_B} \frac{x_B^2}{2} + \xi_B T_C x_B \right] - k_B' (T - T_C) \tag{12.32}$$

in metal B and

$$\frac{x_A}{A_A}E_A = \frac{I}{A_A}\left[\frac{-I}{\kappa_B A_B}\ell_B x_A - \frac{I}{\kappa_A A_A}\frac{x_A^2}{2} - \xi_B(T_H - T_C)x_A + \xi_A T_H x_A\right]$$
$$- k_A'(T - T_H) \tag{12.33}$$

in metal A.

We can now evaluate E_A and E_B by setting $x_A = \ell_A$ and $x_B = \ell_B$. Then

$$Q_H = \left(\frac{k_A' A_A}{\ell_A} + \frac{k_B' A_B}{\ell_B}\right)(T_H - T_C) + IT_H(\xi_A - \xi_B) - \frac{I^2}{2}\left(\frac{\ell_A}{\kappa_A A_A} + \frac{\ell_B}{\kappa_B A_B}\right) \tag{12.34}$$

and

$$Q_C = -\left(\frac{k_A' A_A}{\ell_A} + \frac{k_B' A_B}{\ell_B}\right)(T_H - T_C) - IT_C(\xi_A - \xi_B) - \frac{I^2}{2}\left(\frac{\ell_A}{\kappa_A A_A} + \frac{\ell_B}{\kappa_B A_B}\right). \tag{12.35}$$

The first terms on the right in these equations can be identified as the Fourier heat. The second terms are the Peltier heats, generated or absorbed at the appropriate junctions. The third terms show that the Joule dissipation flows equally to the two heat sinks for the symmetry of the present system.

The above analysis can be applied equally well to a thermoelectric refrigerator. A reversal of the current will reverse the Peltier terms, and this may be sufficient to remove heat from the cold reservoir. In this application, we would usually expect the electrical connection to be made at the higher temperature.

The assumptions in this example can be relaxed, but then the solution procedure must be modified. For a given current I, Eq. 12.28 can be solved for each wire as a second-order differential equation in temperature, with temperature-dependent physical properties and a cross section A dependent on position. The boundary conditions are the set temperatures T_H and T_C at the ends of the wires. Evaluation of Eq. 12.25 now yields the potential of the device.

12.2.6 Optimization

We now wish to complete the design of a thermoelectric heat engine by making a choice of some of the parameters at our disposal. As in the discussion of heat engines in thermodynamics, we shall assume that the temperatures T_H and T_C of our heat reservoirs are fixed, and we shall try to achieve a high value of the efficiency η, defined as the fraction of the heat Q_H, which is actually converted into work:

$$\eta = \frac{VI}{Q_H}. \tag{12.36}$$

We recognize that we can increase both the work and the heat flow by increasing the cross-sectional areas of the wires or by decreasing their lengths, but these scale-dependent aspects are not of present interest.

Instead we ask first how we should set the current I to maximize the efficiency and second how we should select the wire area A_A to obtain the highest efficiency when the current is also set equal to its optimum value. With these two parameters optimized, we may finally examine how the efficiency depends on physical properties such as thermal and electric conductivities and the thermoelectric coefficients.

To optimize the current, we use the shorthand notation

$$R = \frac{\ell_A}{\kappa_A A_A} + \frac{\ell_B}{\kappa_B A_B} \tag{12.37}$$

for the electric resistance and

$$K = \frac{k'_A A_A}{\ell_A} + \frac{k'_B A_B}{\ell_B} \tag{12.38}$$

for the thermal conductance, and we set $\partial\eta/\partial I$ equal to zero to obtain the result

$$Q_H(\Delta\xi\Delta T - 2IR) = VI(T_H\Delta\xi - IR), \tag{12.39}$$

where $\Delta\xi = \xi_A - \xi_B$ and $\Delta T = T_H - T_C$. After simplification, this equation turns out to be quadratic in I, and the optimum current is one of the roots

$$I = \frac{2K\Delta T}{\Delta\xi}\frac{-1 \pm \sqrt{1+Q}}{T_H + T_C}, \tag{12.40}$$

where

$$Q = \frac{(\Delta\xi)^2}{2RK}(T_H + T_C). \tag{12.41}$$

Equation 12.25 tells us that I should be positive if ΔT and $\Delta\xi$ are positive. Hence, we choose the plus sign in Eq. 12.40 and write the optimum current density as

$$I = \frac{\Delta T \Delta\xi / R}{1 + \sqrt{1+Q}}. \tag{12.42}$$

This is below the power maximum inasmuch as the denominator is greater than 2.

To optimize the area ratio, we set $\partial\eta/\partial A_A$ equal to zero and obtain the result

$$Q_H \frac{I^2 \ell_A}{\kappa_A A_A^2} = VI\left(\frac{k_A'}{\ell_A}\Delta T + \frac{I^2 \ell_A}{2\kappa_A A_A^2}\right). \tag{12.43}$$

Solving this in conjunction with Eqs. 12.39 and 12.42 gives the remarkably simple result

$$\frac{A_A}{\ell_A} = \frac{A_B}{\ell_B}\left(\frac{\kappa_B k_B'}{\kappa_A k_A'}\right)^{1/2}. \tag{12.44}$$

When the thermoelectric heat engine is operated at the optimum current and with the optimum area ratio, the efficiency can be expressed as

$$\eta_{\text{opt}} = \frac{\Delta T^*}{\frac{1}{2}\Delta T^* + \left(1 + \sqrt{1+Q}\right)^2} \tag{12.45}$$

where

$$\Delta T^* = \frac{\Delta T(\Delta\xi)^2}{\left(\sqrt{k_A'/\kappa_A} + \sqrt{k_B'/\kappa_B}\right)^2} \tag{12.46}$$

and Q now takes the form

$$Q = \left(\frac{1}{\eta_C} - \frac{1}{2}\right)\Delta T^*, \tag{12.47}$$

η_C being the Carnot efficiency equal to $1 - T_C/T_H$. Figure 12.3 is a graph of these relations. For given temperatures, the efficiency will approach the Carnot efficiency as one finds materials with a higher difference in thermoelectric coefficients or lower ratios of thermal conductivity to electrical conductivity.

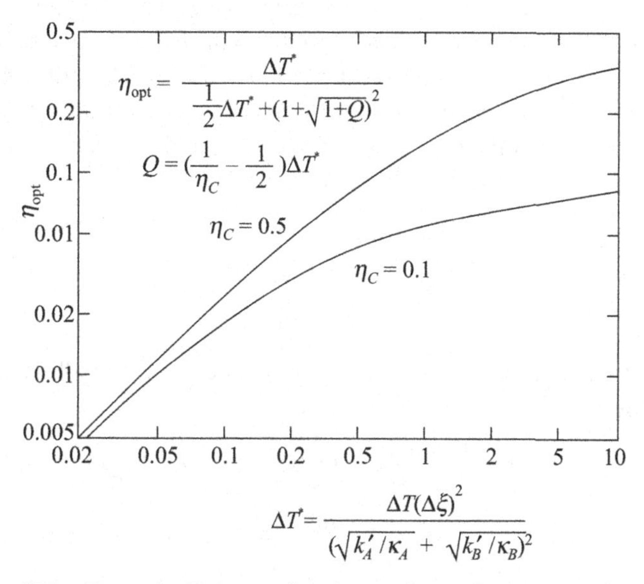

Figure 12.3 Thermal efficiency of a thermoelectric heat engine for which the current and the ratio of wire cross sections have already been given their optimum values.

12.3 Fluctuations and Microscopic Reversibility

Consider a number of essentially identical, isolated slabs containing an ideal-gas mixture of components A and B. These slabs are of length ℓ and are subject to random fluctuations in the local temperature and composition. We shall let the temperature and mole-fraction distributions be represented as Fourier series:

$$T = T_o + \sum_{n=1}^{\infty} a_n \cos\left(\frac{n\pi x}{\ell}\right) \tag{12.48}$$

and

$$y_A = 1 - y_B = y_{Ao} + \sum_{n=1}^{\infty} b_n \cos\left(\frac{n\pi x}{\ell}\right). \tag{12.49}$$

These expressions emphasize that fluctuations can be of macroscopic dimensions and have macroscopic relaxation times. If the convergence of the series is in doubt, they can be truncated after, say, 10 terms. Actually, only moments of energy and mass about the middle of the slab need to be analyzed. It will also become clear that the three-dimensional nature of the fluctuations does not require treatment. We shall assume that the total concentration c and the pressure p are uniform throughout the slab, and we take the molar heat capacities of the pure components to be

$$\tilde{C}_{pA}^{*} = \tilde{C}_{pB}^{*} = 7R/2. \tag{12.50}$$

The student should show that the internal energy is

$$U = (C - nR)T_o, \tag{12.51}$$

a constant, and that if deviations from T_o and y_{Ao} are small, then the entropy is

$$S = -nR\ln p + C\ln T_o - Rn\left[y_{Ao}\ln y_{Ao} + (1 - y_{Ao})\ln(1 - y_{Ao})\right]$$

$$-\sum_{n=1}^{\infty}\frac{Ca_n^2}{4T_o^2} - \frac{Rn}{4y_{Ao}(1-y_{Ao})}\sum_{n=1}^{\infty}b_n^2 + \cdots, \tag{12.52}$$

where n is the total number of moles and $C = 7nR/2$ is the total heat capacity. Show also that the total number of moles of component A is fixed, so that each slab can be regarded to be isolated.

It follows that the entropy-production rate is

$$\frac{dS}{dt} = -\sum_{n=1}^{\infty}\frac{Ca_n}{2T_o^2}\frac{da_n}{dt} - \frac{Rn}{2y_{Ao}(1-y_{Ao})}\sum_{n=1}^{\infty}b_n\frac{db_n}{dt}, \tag{12.53}$$

if deviations from the uniform, average conditions are small.

12.3.1 Macroscopic Part

An important assumption to be introduced later is that fluctuations decay or regress, on the average, in the same way that macroscopic variations in temperature and composition are observed to disappear, leaving a uniform state. We, therefore, want to analyze the temperature and composition profiles (Eqs. 12.48 and 12.49) first according to macroscopic laws.

Under our assumptions, the molar average velocity is zero, and Eqs. 4.2 and 6.3 reduce to[†]

$$\rho\hat{C}\frac{\partial T}{\partial t} = -\frac{\partial q_x'}{\partial x} \tag{12.54}$$

and

$$c\frac{\partial y_A}{\partial t} = -\frac{\partial N_{Ax}}{\partial x}. \tag{12.55}$$

Since we expect effects to be local, we express the fluxes in terms of the gradients as

$$q_x' = -k\frac{\partial T}{\partial x} - L_{qy}\frac{\partial y_A}{\partial x} \tag{12.56}$$

and

$$N_{Ax} = -cD\frac{\partial y_A}{\partial x} - L_{yq}\frac{\partial T}{\partial x}. \tag{12.57}$$

These are equivalent to Eqs. 11.9 and 11.10 if

$$RT_o^2 L_{yq} = y_{Ao}(1-y_{Ao})L_{qy} = c_A y_B RT\left(\frac{D_A^T}{\rho_A} - \frac{D_B^T}{\rho_B}\right), \tag{12.58}$$

and $D = \mathcal{D}_{AB}$. However, we want to prove this symmetry relation rather than assume it.

Now substitute the assumed temperature and mole-fraction distributions (Eqs. 12.48 and 12.49) into Eqs. 12.54–12.57 (taking k, D, L_{qy}, and L_{yq} to be constant), multiply by cos $(m\pi x/\ell)$, and integrate from 0 to ℓ to obtain

$$\rho\hat{C}\frac{da_n}{dt} = -\frac{n^2\pi^2}{\ell^2}(ka_n + L_{qy}b_n) \tag{12.59}$$

and

$$c\frac{db_n}{dt} = -\frac{n^2\pi^2}{\ell^2}(cDb_n + L_{yq}a_n). \tag{12.60}$$

(The readers may wish to show that the second law of thermodynamics implies that

---

[†]Equation 4.2 should perhaps be put into the equivalent form

$$\frac{\partial\rho\hat{H}}{\partial t} + \nabla\cdot(\rho\hat{H}\mathbf{v}) = \frac{Dp}{Dt} - \nabla\cdot\mathbf{q} - \boldsymbol{\tau}:\nabla\mathbf{v}$$

before assuming p to be constant (and neglecting viscous dissipation). See also Problem 4.1.

$$\frac{1}{4}[RT_o^2 L_{yq} + y_{Ao}(1-y_{Ao})L_{qy}]^2 \leq c y_{Ao}(1-y_{Ao})kDRT_o^2. \quad (12.61)$$

This is equivalent to the relation 11.12, but without any assumption about the symmetry of the transport coefficients.)

Solve the differential Eqs. 12.59 and 12.60 with respect to time to obtain

$$a_n = \exp\left\{-\frac{n^2\pi^2}{\ell^2}\frac{\alpha+D}{2}t\right\}\times$$

$$\left[a_n^o \cosh\left(\frac{n^2\pi^2}{2\ell^2}Zt\right) + \frac{a_n^o(D-\alpha) - 2b_n^o L_{qy}/\rho\hat{C}}{Z}\sinh\left(\frac{n^2\pi^2}{2\ell^2}Zt\right)\right] \quad (12.62)$$

and

$$b_n = \exp\left\{-\frac{n^2\pi^2}{\ell^2}\frac{\alpha+D}{2}t\right\}\times$$

$$\left[b_n^o \cosh\left(\frac{n^2\pi^2}{2\ell^2}Zt\right) + \frac{b_n^o(\alpha-D) - 2a_n^o L_{yq}/c}{Z}\sinh\left(\frac{n^2\pi^2}{2\ell^2}Zt\right)\right], \quad (12.63)$$

where a_n^o and b_n^o are the initial values of a_n and b_n, $\alpha = k/\rho\hat{C}$ is the thermal diffusivity, and

$$Z = \sqrt{(\alpha-D)^2 + \frac{4L_{qy}L_{yq}}{\rho\hat{C}c}}. \quad (12.64)$$

12.3.2 Ensemble Averages

We shall use triangular brackets to denote an average over the slabs in the ensemble, for example, $\langle a_n \rangle$. Frequently, it is assumed that the same result would be obtained by averaging over time for a single slab. This assumption is known as the ergodic hypothesis. In problems where average conditions change with time, the ensemble average is safer. The ensemble average is also referred to as the expected value of the quantity being averaged; if an experiment were carried out many times on systems that are identical in their macroscopic aspects, the average of the results would be the ensemble average.

Our slabs are at equilibrium, so that $\langle a_n(t) \rangle = 0$ and $\langle b_n(t) \rangle = 0$. We should like to pick from our ensemble those cases where $a_n(t)$

has given value a_n', say 10^{-15} deg. From those cases alone, we might then look at the expected value $<a_m(t + \tau)>$ or $<b_m(t + \tau)>$ of one of the coefficients at a time $t + \tau$, which is before or after the time t. This is awkward since the number of cases where $a_n(t)$ is exactly 10^{-15} deg should be zero in a finite ensemble. Furthermore, the precise value chosen, $a_n' = 10^{-15}$ deg, has no particular significance to us. We accomplish much the same goal, but on a firmer statistical basis, by focusing our attention on the *correlation* $<a_n(t)a_m(t + \tau)>$ or $<a_n(t)b_m(t + \tau)>$. Thus, $<a_n(t)a_m(t + \tau)>$ can be obtained by taking the averages $<a_m(t + \tau)>$ for particular values of a_n', multiplying by a_n', and averaging over the entire ensemble.

$$< a_n(t)a_m(t+\tau) > = < a_n' < a_m(t+\tau) >_{a_n(t)=a_n'} >. \qquad (12.65)$$

When we use the expression $<c_n(t)a_m(t + \tau)>$ or $<c_n(t)b_m(t + \tau)>$, c_n can refer to either a_n or b_n.

12.3.3 Microscopic Reversibility and Probability of States

On a microscopic scale, all collisions are reversible—at least, in the absence of external fields such as magnetic fields. This is true whether we use quantum mechanics or classical mechanics. If the velocities of all the particles are reversed at an instant, the particle positions retrace, on the average, the positions they traversed before that instant. It is also plausible to assume that such pairs of states are equally probable in the ensemble since they have the same general characteristics. It follows from the principle of microscopic reversibility and the hypothesis of equal probability of occupancy of states that

$$<a_n(t)a_m(t + \tau)> = <a_n(t)a_m(t - \tau)>. \qquad (12.66)$$

This means that the correlation decays both forward in time and backward in time (see Fig. 12.4) in an aged ensemble where each system is at equilibrium. Equation 12.66 also depends on the fact that the Fourier coefficients are even functions of the molecular velocities.

To help visualize Eq. 12.66, pair up in your mind the systems in the ensemble before making the averages. For a system having a value $a_n(t)$, find another system with the same molecular positions but reversed velocities at time t. It is clear that if the first system has the value a_m at time $t + \tau$, then the second system must have had the same value a_m at time $t - \tau$. Conversely, the second system has the same value of a_m at $t + \tau$ that the first system had at $t - \tau$. This pair thus contributes equally to the averages on both sides of Eq. 12.66. This is the meaning of the principle of microscopic reversibility. The hypothesis of equal probability of occupancy of states says that when we have exhausted the pairs in the ensemble, the remaining systems will contribute negligibly to the averages. Equation 12.66 can then be satisfied with any desired accuracy, provided that there are enough systems in the ensemble.

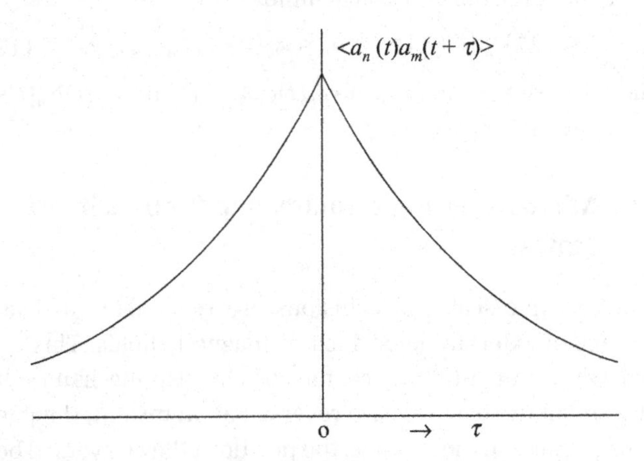

Figure 12.4 Decay of a correlation of fluctuations, with the correlation time τ.

The demonstration of Eq. 12.66 does not require the fluctuations from the equilibrium, average conditions to be small.

Since the correlations are independent of τ in an aged ensemble, Eq. 12.66 can be rewritten without any additional assumptions as

$$\langle a_n(t)a_m(t+\tau)\rangle = \langle a_m(t)a_n(t+\tau)\rangle. \tag{12.67}$$

The arguments are identical for other correlations. Analogous to Eq. 12.67, we obtain

$$\langle a_n(t)b_m(t+\tau)\rangle = \langle b_m(t)a_n(t+\tau)\rangle \tag{12.68}$$

and

$$\langle b_n(t)b_m(t+\tau)\rangle = \langle b_m(t)b_n(t+\tau)\rangle. \tag{12.69}$$

For a microcanonical ensemble in statistical mechanics, the probability of occupancy of a state is proportional to exp (S/k), where k is the Boltzmann constant. This means that the probability of finding a fluctuation with a_1 between a_1 and $a_1 + da_1$, and so forth for $a_2, a_3, ..., b_1, b_2, ...$ is

$$A\, e^{S/K} da_1 da_2 \ldots db_1 db_2 \ldots,$$

where A is a normalization constant. Limit n to 10 so that we do not have to worry about whether all the integrals converge.

The entropy of fluctuations close to equilibrium is given by Eq. 12.52. On this basis, the student can show that most correlations are zero at $\tau = 0$. In particular,

$$<a_n(t)a_m(t)> = 0 \quad \text{if} \quad n \neq m, \tag{12.70}$$

$$<a_n(t)b_m(t)> = 0, \tag{12.71}$$

and

$$<b_n(t)b_m(t)> = 0 \quad \text{if} \quad n \neq m. \tag{12.72}$$

The only nonzero correlations at $\tau = 0$ are

$$< a_n(t)a_n(t) > = \frac{2kT_o^2}{C} \tag{12.73}$$

and

$$< b_n(t)b_n(t) > = \frac{2y_{Ao}(1 - y_{Ao})k}{Rn}. \tag{12.74}$$

12.3.4 Decay of Fluctuations

The last major hypothesis is that fluctuations decay like macroscopic variations. This too is plausible since the mechanism of approach to equilibrium is basically the same random process that leads to fluctuations and their decay. In one interpretation of this hypothesis, we assume that a_n and b_n on the left sides of Eqs. 12.62 and 12.63 are to be replaced by $<c_m(t)a_n(t + \tau)>$ and $<c_m(t)b_n(t + \tau)>$, and that a_n^o and b_n^o on the right sides of these equations are to be replaced by $<c_m(t)a_n(t)>$ and $<c_m(t)b_n(t)>$. Finally, t is to be replaced by $|\tau|$, the absolute value being used since fluctuations decay both forward and backward in time (see Eq. 12.66 and Fig. 12.4).

The right sides of the resulting equations are observed to be zero unless $m = n$ in view of Eqs. 12.70 to 12.72. Thus, all the correlations in Eqs. 12.67 to 12.69 are found to be zero unless $m = n$. Different orders in the Fourier series are not correlated.

If, in the replacement of Eq. 12.62, one sets $c_m(t)$ equal to $a_n(t)$, one obtains the expression for the decay of the correlation $<a_n(t)a_n(t+\tau)>$:

$$<a_n(t)a_n(t+\tau)> = \frac{2kT_o^2}{C}\exp\left\{-\frac{n^2\pi^2}{\ell^2}\frac{\alpha+D}{2}|\tau|\right\}$$

$$\times\left[\cosh\left(\frac{n^2\pi^2}{2\ell^2}Z|\tau|\right)+\frac{D-\alpha}{Z}\sinh\left(\frac{n^2\pi^2}{2\ell^2}Z|\tau|\right)\right]. \quad (12.75)$$

Here Eqs. 12.71 and 12.73 have been used. Equation 12.75 is a guide to the behavior depicted in Fig. 12.4. A similar result is obtained for the decay of the correlation $<b_n(t)b_n(t+\tau)>$ if, in the replacement of Eq. 12.63, one sets $c_m(t)$ equal to $b_n(t)$.

Finally, in the replacement of Eq. 12.62, set $c_m(t)$ equal to $b_n(t)$, and in the replacement of Eq. 12.63, set $c_m(t)$ equal to $a_n(t)$. With Eqs. 12.71, 12.73, and 12.74, we obtain

$$<b_n(t)a_n(t+\tau)> =$$

$$-\frac{4L_{qy}y_{Ao}(1-y_{Ao})k}{Z\rho\hat{C}Rn}\exp\left\{-\frac{n^2\pi^2}{\ell^2}\frac{\alpha+D}{2}|\tau|\right\}\sinh\left(\frac{n^2\pi^2}{2\ell^2}Z|\tau|\right) \quad (12.76)$$

and

$$<a_n(t)b_n(t+\tau)> =$$

$$-\frac{4L_{yq}kT_o^2}{ZcC}\exp\left\{-\frac{n^2\pi^2}{\ell^2}\frac{\alpha+D}{2}|\tau|\right\}\sinh\left(\frac{n^2\pi^2}{2\ell^2}Z|\tau|\right). \quad (12.77)$$

These two expressions are required to be equal by Eq. 12.68. This thus completes the demonstration of the Onsager reciprocal relation 12.58.

12.3.5 Summary

The Onsager reciprocal relations depend on several fundamental principles and hypotheses. Each is plausible by itself, and several are commonly accepted in statistical mechanics or in macroscopic transport theory. Each requires some development for the student so as to reduce its level of abstraction. Together they lead to the desired result. In review, these principles are as follows:

1. Macroscopic transport laws—a linear dependence of diffusion and heat fluxes on the gradients in the system. The relation we want to prove involves transport coefficients so defined.
2. Microscopic reversibility. Molecular collisions proceed equally well in the forward and reverse directions.
3. Statistical distribution of fluctuations. In the analysis of the equilibrium state, statistical mechanics deals with the probability of encountering departures from equilibrium properties.
4. Similar modes of decay of microscopic and macroscopic departures from equilibrium. Both processes are statistical in nature, and there is no good reason for them to be different.

The Onsager reciprocal relations are valuable because they are general and provide equalities among macroscopic quantities without any dependence on the molecular properties of the materials. For this reason, they belong in the study of transport phenomena. On the other hand, the Chapman–Enskog kinetic theory of gases had already led to thermal diffusion and the Dufour effect, with the same transport properties D_i^T appearing in both, but the theory was restricted to dilute mixtures of monatomic gases.

References

1. Lars Onsager. "Reciprocal Relations in Irreversible Processes. I." *Physical Review*, **37**, 405–426 (1931a).
2. Lars Onsager. "Reciprocal Relations in Irreversible Processes. II." *Physical Review*, **38**, 2265–2279 (1931b).
3. Eugene P. Wigner. "Derivations of Onsager's Reciprocal Relations." *The Journal of Chemical Physics*, **22**, 1912–1915 (1954).
4. Charles W. Monroe and John Newman, "Onsager's shortcut to proper forces and fluxes." *Chemical Engineering Science*, **64** (2009), 4804–4809.
5. Bernard D. Coleman and Clifford Truesdell. "On the Reciprocal Relations of Onsager." *The Journal of Chemical Physics*, **33**, 28–31 (1960).

6. James Wei. "Irreversible Thermodynamics in Engineering." *Industrial and Engineering Chemistry*, **58**(10), 55–60 (1966).
7. R. Clerk Jones and W. H. Furry. "The Separation of Isotopes by Thermal Diffusion." *Reviews of Modern Physics*, **18**, 151–224 (1946).
8. John Newman and Nitash P. Balsara. *Electrochemical Systems*, 4th Edition. Hoboken, New Jersey: John Wiley and Sons, 2020.

Suggested Reading for Basic Transport Relations

1. In general, Chapters 3, 11, and 19 of R. Byron Bird, Warren E. Stewart, and Edwin N. Lightfoot, *Transport Phenomena*, 2nd Edition. New York: John Wiley & Sons, 2002. (BSL).
2. Fluid mechanics.
   BSL, 2nd Edition, pp. 11–23 and 77–86.
   Hermann Schlichting, *Boundary-Layer Theory*, 7th Edition, translated by J. Kestin. New York: McGraw-Hill Book Company, 1979, pp. 47–69.
3. Heat transfer in a pure fluid.
   BSL, 2nd Edition, pp. 333–341 and 265–273.
   Hermann Schlichting, *Boundary-Layer Theory*, 7th Edition, translated by J. Kestin. New York: McGraw-Hill Book Company, 1979, pp. 265–276.
4. Mass transfer. Simplified flux equation.
   $\mathbf{J}_A^* = -\mathcal{D}_{AB} c \nabla x_A$
   BSL, 2nd Edition, pp. 513–524 and 582–592.
   Veniamin G. Levich, *Physicochemical Hydrodynamics*. Englewood Cliffs, N. J.: Prentice-Hall, Inc., 1962, pp. 39–56.
5. Multicomponent diffusion equation.
   $$c_i \nabla \mu_i = RT \sum_j \frac{c_i c_j}{c\mathcal{D}_{ij}} (\mathbf{v}_j - \mathbf{v}_i)$$
   BSL, 2nd Edition, p. 538.
   E. N. Lightfoot, E. L. Cussler, Jr., and R. L. Rettig, "Applicability of the Stefan–Maxwell equations to multicomponent diffusion in liquids," *AIChE Journal*, **8**, 708–710, 1962.
   C. Truesdell, "Mechanical Basis of Diffusion," *The Journal of Chemical Physics*, **37**, 2336–2344, 1962.
6. Heat transfer in a mixture. BSL, 2nd Edition, pp. 590–592.
7. Definition of transport properties and their behavior as functions of temperature, pressure, and composition for gases and liquids.
   BSL, 2nd Edition, pp. 21–31, 272–280, 521–531, and 536–538.
   Sydney Chapman and T. G. Cowling, *The Mathematical Theory of Non-Uniform Gases*. Cambridge University Press, 1970.

Bruce E. Poling, John M. Prausnitz, and John P. O'Connell, *The Properties of Gases and Liquids*. New York: McGraw-Hill Book Company, 2001.

Notation

$\mathbf{a}$	acceleration
c	total molar concentration
c_i	molar species concentration
C_V	heat capacity at constant volume
d	diameter of hard sphere
D_{ij}	diffusion coefficient
$\mathcal{D}_{ij}$	diffusion coefficient
D_i^T	thermal diffusion coefficient
$\mathbf{e}$	total energy flux
e	magnitude of the electronic charge
f	velocity distribution function
F	Faraday's constant
$\mathbf{F}$	force
g	entropy production per unit volume
$\mathbf{g}$	acceleration due to gravity
G	Gibbs free energy
H	enthalpy
ima	intermolecular attraction
imf	intermolecular force
$\mathbf{j}_i$	mass flux of species i with respect to the mass average velocity
$\mathbf{J}_i$	molar flux of species i with respect to the mass average velocity
$\mathbf{J}_i^*$	molar flux of species i with respect to the molar average velocity
k	thermal conductivity
$\widehat{K}$	total energy per unit mass
L	Loschmidt number or Avogadro's number
m	mass
M_{i}	molar mass
n	number of species
n	number density

n_i	number of moles of species *i*
$\mathbf{n}_i$	mass flux of species *i*
$\mathbf{N}_i$	molar flux of species *i*
p	pressure
$\mathbf{q}$	heat flux
$\mathbf{q}'$	see Eq. 11.3
r_i	rate of production of species *i*
R_i	rate of production of species *i*
R	gas constant
S	entropy
t	time
T	temperature
$\bar{u}$	average speed of particles due to thermal motion
$\hat{U}$	internal energy per unit mass
$\mathbf{v}$	(mass) average velocity
$\mathbf{v}^*$	molar average velocity
$\mathbf{v}_i$	average velocity of species *i*
V	volume
x_i	mole fraction of species *i*
x,y,z	spatial coordinates
ϵ	average internal (rotational and vibrational) energy of molecules
ϵ	Lennard–Jones energy parameter
$\mathbf{I}$	unit tensor
κ	bulk viscosity
κ	Boltzmann constant
λ	mean free path
μ	viscosity
μ_i	chemical potential of species *i*
ν	kinematic viscosity
$\boldsymbol{\nu}$	dummy velocity
ρ	density
σ	electric conductivity
σ	Lennard–Jones distance parameter
$\boldsymbol{\sigma}$	entropy flux
$\boldsymbol{\tau}$	stress or viscous momentum flux

$\boldsymbol{\phi}$	total momentum flux
ϕ	intermolecular potential energy function
Φ	electrostatic potential
Φ_v	dissipation function
ω_i	mass fraction of species *i*
Ω	function for Lennard–Jones treatment of transport properties
^	per unit mass
*	transpose
i	species
$\bar{i}$	partial molal

SECTION B:
LAMINAR FLOW SOLUTIONS

Chapter 13

Introduction

In Chapters 13 through 30, we shall study laminar flow solutions for problems arising in fluid mechanics and mass transfer. The preceding chapters provide a complicated set of nonlinear partial differential equations, which presumably describe such processes but which are quite difficult to solve. Simplifying assumptions are usually necessary.

In these notes, we shall, for the most part, assume that all fluid properties are constant. Thus, for an incompressible Newtonian fluid, the stress relation 2.3 may be written as

$$\boldsymbol{\tau} = -\mu[\nabla \mathbf{v} + (\nabla \mathbf{v})^*], \tag{13.1}$$

and the equations of continuity 2.1 and motion 2.2 become

$$\boxed{\nabla \cdot \mathbf{v} = 0.} \tag{13.2}$$

$$\boxed{\rho\left(\frac{\partial \mathbf{v}}{\partial t} + \mathbf{v} \cdot \nabla \mathbf{v}\right) = -\nabla p + \mu \nabla^2 \mathbf{v} + \rho \mathbf{g}.} \tag{13.3}$$

(Note that $\nabla \cdot (\nabla \mathbf{v})^* = 0$ for an incompressible fluid.)

Furthermore, for an incompressible fluid with no free surfaces, it is frequently convenient to combine the pressure term and the gravitational term:

$$-\nabla \mathcal{P} = -\nabla p + \rho \mathbf{g}. \tag{13.4}$$

The Newman Lectures on Transport Phenomena
John Newman and Vincent Battaglia

ISBN 978-981-4774-27-7 (Hardcover), 978-1-315-10829-2 (eBook)
www.jennystanford.com

This is permissible (for an incompressible fluid) as long as **g** can be expressed as $-\nabla\Phi$, the gradient of a gravitational potential. Then the equation of motion becomes

$$\boxed{\rho\left(\frac{\partial \mathbf{v}}{\partial t}+\mathbf{v}\cdot\nabla\mathbf{v}\right)=-\nabla\mathcal{P}+\mu\nabla^2\mathbf{v}.} \tag{13.5}$$

For a fluid of constant viscosity and density, variations of pressure are only of dynamical significance, serving to accelerate or decelerate the fluid but not to cause the fluid properties to vary. The hydrostatic pressure gradient $\rho\mathbf{g}$ does not contribute significantly to the accelerating force, and Eq. 13.4 can be regarded as subtracting the hydrostatic pressure from the total or thermodynamic pressure p to leave the "dynamic pressure" $\mathcal{P}$.

For over a century, fluid mechanists have diligently sought solutions to Eqs. 13.2 and 13.3 or Eqs. 13.2 and 13.5.

We shall frequently assume that mass transfer processes are described by the equation of convective diffusion:

$$\boxed{\frac{\partial c_i}{\partial t}+\mathbf{v}\cdot\nabla c_i=D\nabla^2 c_i.} \tag{13.6}$$

Here c_i represents the concentration of the diffusing species and D is a diffusion coefficient, assumed to be constant. This equation can describe diffusion in a binary mixture under certain conditions. If we substitute Eq. 6.5 into Eq. 6.3 and assume that ρ and $\mathcal{D}_{AB}$ are constants and that there are no reactions in the bulk of the solution ($r_A = 0$), then we obtain

$$\frac{\partial \rho_A}{\partial t}+\mathbf{v}\cdot\nabla\rho_A=\mathcal{D}_{AB}\nabla^2\rho_A.$$

Division by the molar mass gives

$$\frac{\partial c_A}{\partial t}+\mathbf{v}\cdot\nabla c_A=\mathcal{D}_{AB}\nabla^2 c_A,$$

which is the same as Eq. 13.6. Under certain conditions, Eq. 13.6 can describe transport in multicomponent systems (see Problem 8.4).

For heat transfer in a pure fluid of constant density and thermal conductivity, Eqs. 4.3 and 4.4 yield

$$\rho\hat{C}_V\left(\frac{\partial T}{\partial t}+\mathbf{v}\cdot\nabla T\right)=k\nabla^2 T-\boldsymbol{\tau}:\nabla\mathbf{v}.$$

If $\hat{C}_V$ can be taken to be constant and if viscous dissipation can be neglected, this equation is of the same form as the equation of convective diffusion 13.6, and the thermal diffusivity is $k / \rho\hat{C}_V$.† Consequently, we shall not treat heat-transfer problems separately from mass-transfer problems. The presence of an interfacial velocity in mass transfer is one of the differences between heat and mass transfer and will be considered in Chapter 29.

To reiterate, we shall consider hydrodynamic problems described by Eqs. 13.2 and 13.3 and mass-transfer processes described by Eq. 13.6. This means that we are assuming several of the fluid properties to be constant. This can be justified on the basis of the resulting simplification of the problem. Solutions are sometimes obtained where otherwise the situation is hopeless. Furthermore, the variation of properties is peculiar to each particular fluid, and solutions accounting for this variation are of limited interest, whereas constant-property solutions are more general and can frequently apply to both heat and mass transfer with only slight changes of notation. Thus, constant-property solutions have significant pedagogic advantages.

---

†The thermal diffusivity is usually stated to be $k/\rho\hat{C}p$, but for condensed phases, the difference is negligible.

Chapter 14

Simple Flow Solutions

Here we want to mention some relatively simple flow problems. In all these problems, we can eliminate the explicit appearance of the gravitational terra by using the dynamic pressure defined by Eq. 13.4.

14.1 Steady Flow in a Pipe or Poiseuille Flow

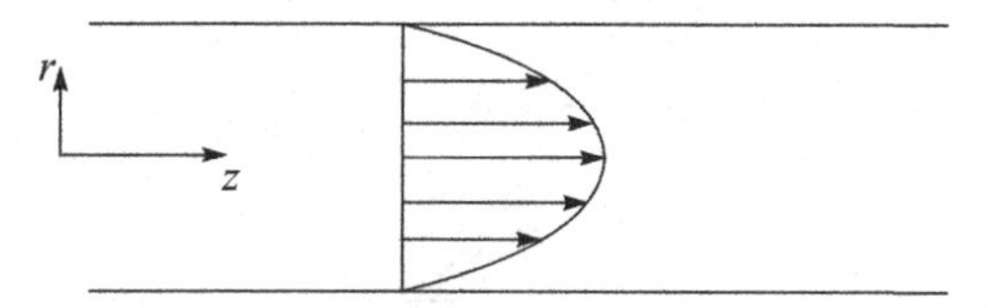

For fully developed, steady flow in a pipe, we expect the velocity profile to be independent of position z along the pipe and also of the angle θ. The equations of motion then yield the familiar result:

$$v_z = -\frac{d\mathcal{P}}{dz}\frac{R^2}{4\mu}\left[1-\left(\frac{r}{R}\right)^2\right], \tag{14.1}$$

where R is the radius of the pipe.

The Newman Lectures on Transport Phenomena
John Newman and Vincent Battaglia

ISBN 978-981-4774-27-7 (Hardcover), 978-1-315-10829-2 (eBook)
www.jennystanford.com

14.2 Couette Flow

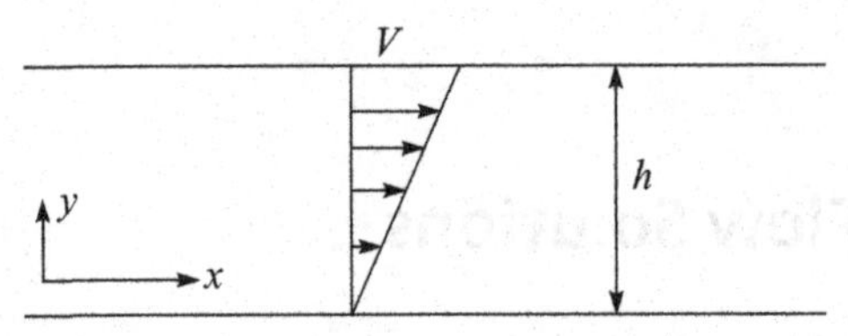

Here the fluid is between two large plates, which are moving past each other with a relative velocity *V*. By the same reasoning as for pipe flow, the velocity profile is

$$v_x = V\frac{y}{h}. \tag{14.2}$$

14.3 Impulsive Motion of a Flat Plate

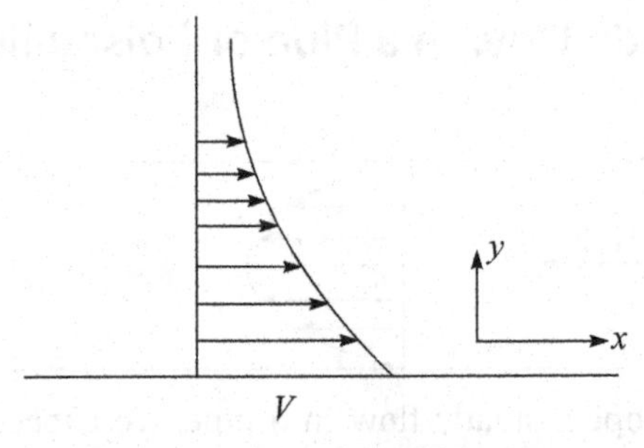

This problem, also known as Stokes first problem, involves the flow in a large, initially stagnant fluid as a consequence of the bounding solid surface being given a constant tangential velocity *V* at time t = 0. In this simple flow system, we expect that $v_y = v_z = 0, \nabla \mathcal{P} = 0$, and that v_x depends only on *t* and *y*. The *x*-component of the equation of motion, with zero terms discarded, is

$$\rho\frac{\partial v_x}{\partial t} = \mu\frac{\partial^2 v_x}{\partial y^2}, \tag{14.3}$$

and the boundary and initial conditions are

$$\left.\begin{aligned} v_x &= 0 \quad \text{at} \quad t = 0. \\ v_x &= 0 \quad \text{at} \quad y = \infty. \\ v_x &= V \quad \text{at} \quad y = 0. \end{aligned}\right\} \tag{14.4}$$

Since the problem is linear, the use of the ratio $\phi = v_x/V$ eliminates the parameter V:

$$\frac{\partial \phi}{\partial t} = \nu \frac{\partial^2 \phi}{\partial y^2}, \quad \begin{array}{l} \phi = 0 \text{ at } t = 0. \\ \phi = 0 \text{ at } y = \infty. \\ \phi = 1 \text{ at } y = 0. \end{array}$$

Now ϕ can depend only on dimensionless combinations of ν, t, and y. But there is only one independent dimensionless group, for which we choose $\eta = y/\sqrt{4\nu t}$, and the velocity ratio ϕ can depend only on η). This *similarity transformation* from the variables y and t to the variable η can be effected by means of the chain rule of differentiation:

$$\frac{\partial \phi}{\partial t} = \frac{d\phi}{d\eta}\frac{\partial \eta}{\partial t} = -\frac{\eta}{2t}\frac{d\phi}{d\eta}.$$

$$\frac{\partial \phi}{\partial y} = \frac{d\phi}{d\eta}\frac{\partial \eta}{\partial y} = \frac{1}{\sqrt{4\nu t}}\frac{d\phi}{d\eta}.$$

$$\frac{\partial^2 \phi}{\partial y^2} = \frac{1}{\sqrt{4\nu t}}\left(\frac{d}{d\eta}\frac{d\phi}{\partial \eta}\right)\frac{\partial \eta}{\partial y} = \frac{1}{4\nu t}\frac{d^2\phi}{d\eta^2}.$$

Substitution into the differential equation gives

$$\frac{\partial^2 \phi}{\partial \eta^2} + 2\eta \frac{d\phi}{d\eta} = 0, \tag{14.5}$$

and the boundary conditions become

$$\left.\begin{array}{l} \phi = 0 \text{ at } \eta = \infty. \\ \phi = 1 \text{ at } \eta = 0. \end{array}\right\} \tag{14.6}$$

The solution to this differential equation is

$$\phi = 1 - \frac{2}{\sqrt{\pi}}\int_0^\eta e^{-\eta^2} d\eta = 1 - \operatorname{erf}(\eta), \tag{14.7}$$

where $\operatorname{erf}(\eta)$ is a tabulated function. The shear stress at the wall is

$$\tau_{yx}\Big|_{y=0} = -\mu \frac{\partial v_x}{\partial y}\Big|_{y=0} = -\frac{\mu V}{\sqrt{4\nu t}}\frac{d\phi}{d\eta}\Big|_{\eta=0} = \rho V\sqrt{\frac{\nu}{\pi t}}. \tag{14.8}$$

The shear stress is infinite at $t = 0$, which is to be expected since the velocity gradient is infinite. Subsequently, the stress decreases like $1/\sqrt{t}$ and is proportional to $\rho V\sqrt{\nu}$. As time proceeds, the disturbance propagates into the fluid. The distance to which it has spread is proportional to $\sqrt{\nu t}$. The velocity profiles are similar in the sense that they can be made to coincide by stretching the y-axis by an amount proportional to $1/\sqrt{t}$.

One should notice how the similarity transformation was used to reduce the partial differential equation 14.3 to the ordinary differential equation 14.5. This was effected by a combination of the variables y and t into one variable η. At the same time, it was necessary for two of the conditions 14.4 to yield just one of the boundary conditions 14.6. Other similarity transformations will be used in subsequent chapters of this book (see also Appendix B), and an understanding of its use here will be helpful. Other useful mathematical techniques are described in Volume 1 of this series. These include coordinate transformations and methods for solving ordinary and partial differential equations.

Problems

14.1 Show how Eqs. 14.5 and 14.6 yield Eq. 14.7. Notice that the dependent variable ϕ does not appear in Eq. 14.5, so that this equation can be regarded as a first order, linear differential equation for $d\phi/d\eta$. Since the variables are separable, one can integrate immediately to get

$$\frac{d\phi}{d\eta} = Ae^{-\eta^2}.$$

14.2 Transient diffusion to a sphere in a stagnant medium satisfies the equation

$$\frac{\partial c_i}{\partial t} = D\nabla^2 c_i = \frac{D}{r^2}\frac{\partial}{\partial r}\left(r^2 \frac{\partial c_i}{\partial r}\right).$$

(a) For the boundary and initial conditions

$c_i = c_0$ at $r = R$, $c_i \to c_\infty$ as $r \to \infty$, $c_i = c_\infty$ at $t = 0$,

show that the transformation

$$\phi = r\frac{c_i - c_\infty}{c_0 - c_\infty}$$

reduces the problem to

$$\frac{\partial \phi}{\partial t} = D\frac{\partial^2 \phi}{\partial r^2}$$

with the conditions

$$\phi = R \text{ at } r = R, \quad \frac{\phi}{r} \to 0 \text{ as } r \to \infty, \quad \phi = 0 \text{ at } t = 0.$$

(b) Solve for the Nusselt number

$$\text{Nu} = \frac{2RN_{\text{avg}}}{D(c_\infty - c_o)}$$

as a function of time, and show that Nu $\to 2$ as $t \to \infty$. Here N_{avg} is the flux to the sphere averaged over the surface.

(c) The solution treated above describes a diffusion layer that grows with time. Since the region of concentration variations is very thin at small times, it is reasonable to suppose that this solution would also be applicable for short times for transient diffusion to a sphere in Stokes flow. Estimate the magnitude of the time at which the effect of convection would become important.

14.3 After nucleation, a bubble grows in a supersaturated solution of concentration c_∞ (or a superheated solution in the case of boiling) by a process of mass transfer by convection and diffusion (or thermal conduction) from the supersaturated region to the surface of the bubble. You are asked to predict the growth-rate law. You may make the following assumptions:

(a) The bubble is spherical and is growing radially. The fluid velocity is only in the radial direction and is due to the radial growth of the bubble. In other words, neglect convection due to the tendency of the bubble to rise in the solution, and neglect stray convective currents.

(b) The temperature is uniform, and the heat of solution can be neglected.

(c) The pressure inside the bubble is constant, and the concentration of the solute gas is at the constant, saturation value c_o.

This means that the pressure difference due to surface tension and the pressure due to the radial growth are neglected.

(d) The concentration differs from the bulk value c_∞ only in a thin diffusion layer near the surface of the bubble. This is probably a good assumption for the low diffusion coefficients encountered in liquid-phase diffusion.

(e) The vapor pressure of the solvent can be neglected, and the bubble is a pure gas of the solute material.

It is probably best to proceed by first expressing the velocity in the solution in terms of the growth rate of the bubble and then introducing approximations for the fluid velocity justified by assumption (d) into the equation of convective diffusion in spherical coordinates after transformation from the variable r to the variable

$$y = r - R(t),$$

the normal distance from the surface of the sphere. Then seek a solution to the resulting partial differential equation by means of a similarity transformation of the form

$$\eta = y/g(t).$$

14.4 Treat steady, fully developed, laminar flow of a nonnewtonian fluid in a pipe. Ignore temperature variations. The stress law can be expressed as (Ellis model)

$$\nabla \mathbf{v} + (\nabla \mathbf{v})^* = -\left[\phi_o + \phi_1 (\boldsymbol{\tau} : \boldsymbol{\tau})^{\frac{\alpha-1}{2}}\right]_{\boldsymbol{\tau}}$$

or

$$\frac{\partial v_z}{\partial v} = -\left(\phi_o + \phi_1 \,|\, \tau_{rz} \,|^{\alpha-1}\right)\tau_{rz}.$$

The parameters for the fluid in question are $\alpha = 1.185$, ϕ_o = 0.4210 cm^2/s-dyne, and $\phi_1 = 0.2724$ $cm^{2\alpha}$/s-dyne$^\alpha$. The fluid is incompressible.

(a) By means of a shell momentum balance or, better and simpler, a momentum balance on a core of fluid of radius r, determine the distribution of shear stress τ_{rz}.

(b) From the shear stress distribution, determine the velocity distribution across the tube.

14.5 For impulsive motion of a large flat plate in a power-law fluid intially at rest and described by

$$\eta = m\left|\frac{1}{2}I_2\right|^{(n-1)/2},$$

show that the boundary-layer thickness grows as $t^{1/(n+1)}$ and that the shear stress at the wall decreases as $t^{-n/(n+1)}$, where t is time. For the definition of the invariant I_2, see Ref. [1], p. 102.

14.6 The Hagen–Poiseuille law

$$Q = \int_0^R v_z(r) 2\pi r \, dr = \frac{\pi}{8}\frac{\Delta P}{L}\frac{R^4}{\mu}$$

specifies the volumetric flow rate Q for a Newtonian fluid of viscosity μ for steady, laminar flow in a long circular pipe of radius R and a pressure drop of ΔP over a length L. Is the dependence of the volumetric flow rate on tube radius stronger for a power-law fluid or for a Newtonian fluid? Treat the usual case of a shear-thinning fluid ($n < 1$). Show how the dependence arises.

14.7 A solution (or suspension) of 4% paper pulp in water can be treated as a power-law nonnewtonian fluid (see Ref. [2], pp. 241–244). Such a fluid flows past a flat plate at zero incidence under conditions where viscous forces are negligible except in a thin boundary layer near the solid surface.

(a) Formulate a suitable boundary-layer problem for the flow of the fluid past the plate. Be sure to indicate clearly the coordinate system used, the dependent variables sought, the governing equations by which you are going to seek these unknowns, and a proper and complete set of boundary conditions appropriate to this problem.

(b) Develop order-of-magnitude arguments by which you can tell how the boundary-layer thickness 5 depends on the distance from the leading edge of the plate as well as the velocity v_∞ far from the plate and the properties of the fluid.

(c) Develop a similarity transformation by which you reduce the partial differential equation(s) of Part (b) to an ordinary differential equation from which you can extract the velocity profiles. You are not expected to solve such a differential equation, but you should confirm that the conditions for a successful similarity transformation have been met and that the boundary-layer thickness follows a form compatible with the result of Part (b).

References

1. R. Byron Bird, Warren E. Stewart, and Edwin N. Lightfoot. *Transport Phenomena*. New York: John Wiley & Sons, 1960.
2. R. Byron Bird, Warren E. Stewart, and Edwin N. Lightfoot. *Transport Phenomena*, 2nd Edition. New York: John Wiley & Sons, 2002.

Chapter 15

Stokes Flow past a Sphere

Consider steady flow past a sphere of radius R.

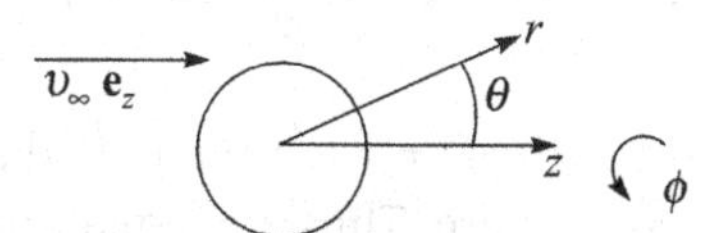

The fluid flow is governed by the Navier–Stokes equation 14.5, and the boundary conditions are

$$\left.\begin{array}{l}\mathbf{v} \rightarrow v_\infty \mathbf{e}_z \\ \mathcal{P} \rightarrow \mathcal{P}_\infty\end{array}\right\} \quad \text{as} \quad r \rightarrow \infty.$$

$$\mathbf{v} = 0 \quad \text{at} \quad r = R.$$

If we introduce dimensionless variables

$$\mathbf{v}^* = \mathbf{v} / v_\infty, \quad \mathbf{r}^* = \mathbf{r} / R, \quad \mathcal{P}^* = \frac{\mathcal{P} - \mathcal{P}_\infty}{\mu v_\infty / R},$$

then the equation of motion becomes (at steady state)

$$\frac{\text{Re}}{2} \mathbf{v}^* \cdot \nabla^* \mathbf{v}^* = -\nabla^* \mathcal{P}^* + \nabla^{*2} \mathbf{v}^*,$$

where Re = $2R\rho v_\infty/\mu$ is the Reynolds number and ∇^* denotes differentiation with respect to the dimensionless coordinates.

Stokes flow deals with flow at low velocities or, more generally, at low Reynolds numbers. At very low values of the Reynolds number,

The Newman Lectures on Transport Phenomena
John Newman and Vincent Battaglia

ISBN 978-981-4774-27-7 (Hardcover), 978-1-315-10829-2 (eBook)
www.jennystanford.com

it seems reasonable to neglect the term on the left altogether. This is what we shall do, and the equation of motion becomes

$$\nabla \mathcal{P} = \mu \nabla^2 \mathbf{v}. \qquad (15.1)$$

The curl of this equation is

$$\nabla \times \nabla \mathcal{P} = 0 = \mu \nabla \times \nabla^2 \mathbf{v}.$$

Since the curl of the gradient of any scalar is always zero, the pressure is thereby eliminated. By means of the vector identity

$$\nabla^2 \mathbf{a} = \nabla \nabla \cdot \mathbf{a} - \nabla \times \nabla \times \mathbf{a}$$

we can write

$$\nabla \times \nabla^2 \mathbf{v} = -\nabla \times \nabla \times \nabla \times \mathbf{v} + \nabla \times \nabla \overbrace{\nabla \cdot \mathbf{v}}^{0} = \nabla^2 \nabla \times \mathbf{v} - \nabla \nabla \cdot \nabla \times \mathbf{v}.$$

Let us introduce the vorticity $\mathbf{\Omega} = \nabla \times \mathbf{v}$ and note that the divergence of the curl of any vector is zero. Thus, the equation to be solved for the vorticity is

$$\nabla^2 \mathbf{\Omega} = 0, \qquad (15.2)$$

which we have obtained by eliminating the pressure from the equation of motion.

Far from the sphere,

$$v_r \to v_\infty \cos\theta \quad \text{and} \quad v_\theta \to -v_\infty \sin\theta. \qquad (15.3)$$

This suggests that we should seek a solution of the form

$$v_r = f(r)\cos\theta, \quad v_\theta = g(r)\sin\theta, \qquad (15.4)$$

where f and g are, as yet, undetermined functions.

In spherical coordinates, the continuity equation 14.2 reads

$$\frac{1}{r^2}\frac{\partial}{\partial r}\left(r^2 v_r\right) + \frac{1}{r\sin\theta}\frac{\partial}{\partial\theta}(v_\theta \sin\theta) = 0,$$

where we have assumed that $v_\phi = 0$. The continuity equation provides a relationship between f and g defined by Eq. 15.4:

$$g = -\frac{1}{2r}\frac{d}{dr}\left(r^2 f\right). \qquad (15.5)$$

For flow past a sphere, where there are no variations with the angle ϕ, the only nonvanishing component of $\mathbf{\Omega}$ is

$$\Omega_\phi = \frac{1}{r}\frac{\partial}{\partial r}(r v_\theta) - \frac{1}{r}\frac{\partial v_r}{\partial\theta} = \frac{\sin\theta}{r}\left[f + \frac{d}{dr}(rg)\right] = h(r)\sin\theta, \qquad (15.6)$$

where

$$h = \frac{1}{r}\left[f + \frac{d}{dr}(rg) \right]. \qquad (15.7)$$

In spherical coordinates, the relevant component of Eq. 15.2 is

$$\frac{1}{r^2}\frac{\partial}{\partial r}\left(r^2 \frac{\partial \Omega_\phi}{\partial r} \right) + \frac{1}{r^2 \sin\theta}\frac{\partial}{\partial \theta}\left(\sin\theta \frac{\partial \Omega_\phi}{\partial \theta} \right) = \frac{\Omega_\phi}{r^2 \sin^2\theta}$$

or

$$\frac{d}{dr}\left(r^2 \frac{dh}{dr} \right) = 2h. \qquad (15.8)$$

This is an Euler differential equation with solutions of the form $h = r^n$. Substitution into the equation shows that $n = 1$ and $n = -2$ are possible values. We discard the solution $h = r$ since $\Omega_\phi \to 0$ as $r \to \infty$. Hence the vorticity is

$$h = \frac{A}{r^2} \quad \text{or} \quad \Omega_\phi = \frac{A}{r^2}\sin\theta, \qquad (15.9)$$

where A is an integration constant.

Substitution of Eq. 15.5 into Eq. 15.7 gives the differential equation for f:

$$\frac{A}{r} = -2r\frac{df}{dr} - \frac{r^2}{2}\frac{d^2 f}{dr^2}. \qquad (15.10)$$

This is also an Euler differential equation, and for homogeneous solutions of the form r^n, we obtain the values $n = 0$ and $n = -3$. Since a particular solution is A/r, the general solution to Eq. 15.10 is

$$f = \frac{A}{r} + C + \frac{D}{r^3}. \qquad (15.11)$$

From the boundary condition (15.3) at $r = \infty$, it is evident that $C = \upsilon_\infty$.

With f given by Eq. 15.11, it follows from Eq. 15.5 that

$$g = -\upsilon_\infty - \frac{A}{2r} + \frac{D}{2r^3}. \qquad (15.12)$$

The boundary conditions $\upsilon_r = \upsilon_\theta = 0$ at $r = R$ can be used to evaluate the constants A and D:

$$D = \frac{1}{2}\upsilon_\infty R^3, \quad A = -\frac{3}{2}\upsilon_\infty R. \qquad (15.13)$$

The above analysis has now yielded the following velocity and vorticity distributions:

$$v_r = v_\infty \left(1 - \frac{3}{2}\frac{R}{r} + \frac{1}{2}\frac{R^3}{r^3}\right)\cos\theta, \tag{15.14}$$

$$v_\theta = -v_\infty \left(1 - \frac{3}{4}\frac{R}{r} - \frac{1}{4}\frac{R^3}{r^3}\right)\sin\theta, \tag{15.15}$$

$$\Omega_\phi = -\frac{3}{2} v_\infty \frac{R}{r^2}\sin\theta. \tag{15.16}$$

The pressure distribution can now be obtained from the equation of motion (15.1) with the result

$$\mathcal{P} = \mathcal{P}_\infty - \frac{3}{2}\frac{\mu R v_\infty}{r^2}\cos\theta. \tag{15.17}$$

One can then calculate the drag (see, for example, Ref. [1], pp. 58–61), which is the famous Stokes law:

$$D = 6\pi\mu v_\infty R. \tag{15.18}$$

The assumption was made in the above analysis that the inertial terms in the equation of motion were small compared to the pressure terms or the viscous terms. It would be instructive to use the above solution to check the validity of this assumption, particularly far from the sphere.

Problems

15.1 In the Stokes analysis of flow past a sphere, it was assumed that the inertial terms in the equation of motion were small compared to the pressure term or the viscous terms. Use the Stokes solution to check this assumption, particularly far from the sphere.

15.2 The velocity profile for flow past a spherical bubble at moderate Reynolds numbers can be approximated as

$$v_r = v_\infty \left(1 - \frac{R^3}{r^3}\right)\cos\theta, \quad v_\theta = -v_\infty \left(1 + \frac{R^3}{2r^3}\right)\sin\theta, \quad v_\phi = 0,$$

where R is the radius of the bubble.

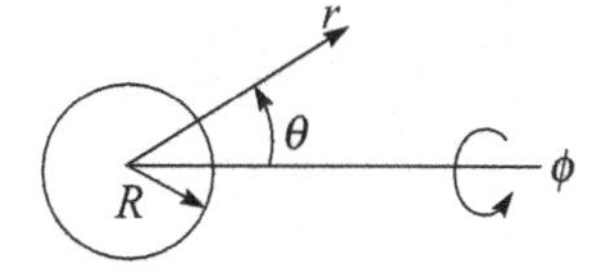

Show that this velocity distribution

(a) corresponds to that for an incompressible fluid.
(b) yields zero for the vorticity.
(c) is an exact solution to the Navier–Stokes equation (for constant viscosity and density). In the course of doing this, you should determine the variation of pressure.

15.3 It is claimed that for the electrophoretic motion of a solid dielectric sphere of radius R, the velocity distribution (in spherical coordinates, with the origin at the center of the particle) is

$$v_r = \left(1 - \frac{R^3}{r^3}\right) v_\infty \cos\theta, \quad v_\theta = -\left(1 + \frac{R^3}{2r^3}\right) v_\infty \sin\theta, \quad v_\phi = 0.$$

(a) What condition is satisfied at infinity, and what condition is satisfied at the surface of the sphere?
(b) Show that the velocity is irrotational, that is, $\nabla \times \mathbf{v} = 0$.
(c) Show that the velocity satisfies the continuity equation for an incompressible fluid.
(d) Show that the velocity is an exact solution to the Navier–Stokes equation for a fluid of constant density and viscosity.
(e) Find an expression for the pressure distribution within the fluid.

Reference

1. R. Byron Bird, Warren E. Stewart, and Edwin N. Lightfoot. *Transport Phenomena*, 2nd Edition. New York: John Wiley & Sons, 2002.

Chapter 16

Flow to a Rotating Disk

We want to study the steady flow of an incompressible fluid caused by the rotation of a large disk about an axis through its center. For this purpose, we use cylindrical coordinates r, θ, and z, where z is the perpendicular distance from the disk and r is the radial distance from the axis of rotation. The velocity on the surface of the disk is

$$v_r = 0, \quad v_z = 0, \quad v_\theta = r\Omega. \tag{16.1}$$

The last condition expresses the fact that the rotating disk drags the adjacent fluid with it at an angular velocity Ω (rad/s). Because of the rotation, there is a centrifugal effect, which tends to throw the fluid out in the radial direction. This will result in a radial component of the velocity that is zero at the surface, has a maximum value near the surface, and then goes to zero again at greater distances from the disk. In order to replace the liquid flowing out in the radial direction, it is necessary to have a z-component of the velocity that brings the fluid toward the disk from far away. This gives us a qualitative picture of the flow field in which none of the velocity components is zero.

The Newman Lectures on Transport Phenomena
John Newman and Vincent Battaglia

ISBN 978-981-4774-27-7 (Hardcover), 978-1-315-10829-2 (eBook)
www.jennystanford.com

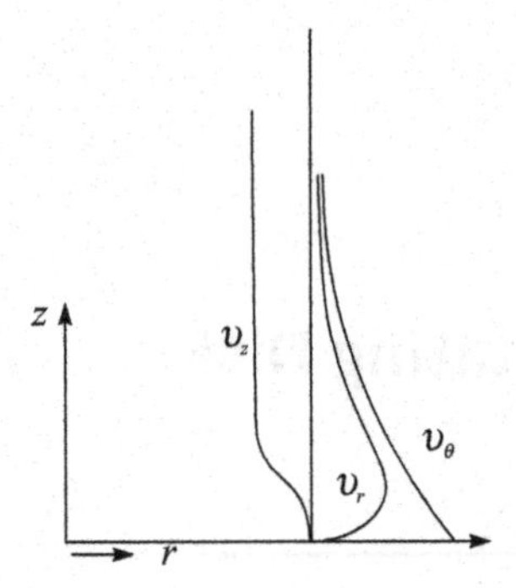

For this flow field with axial symmetry, the equation of continuity 14.2 becomes

$$\frac{1}{r}\frac{\partial}{\partial r}(rv_r)+\frac{\partial v_z}{\partial z}=0 \tag{16.2}$$

and the components of the equation of motion 14.5 are (*r*-component)

$$\rho\left(v_r\frac{\partial v_r}{\partial r}-\frac{v_\theta^2}{r}+v_z\frac{\partial v_r}{\partial z}\right)=-\frac{\partial \mathcal{P}}{\partial r}+\mu\left[\frac{\partial}{\partial r}\left(\frac{1}{r}\frac{\partial}{\partial r}(rv_r)\right)+\frac{\partial^2 v_r}{\partial z^2}\right]. \tag{16.3}$$

(θ-component)

$$\rho\left(v_r\frac{\partial v_\theta}{\partial r}+\frac{v_r v_\theta}{r}+v_z\frac{\partial v_\theta}{\partial z}\right)=\mu\left[\frac{\partial}{\partial r}\left(\frac{1}{r}\frac{\partial}{\partial r}(rv_\theta)\right)+\frac{\partial^2 v_\theta}{\partial z^2}\right]. \tag{16.4}$$

(*z*-component)

$$\rho\left(v_r\frac{\partial v_z}{\partial r}+v_z\frac{\partial v_z}{\partial z}\right)=-\frac{\partial \mathcal{P}}{\partial z}+\mu\left[\frac{1}{r}\frac{\partial}{\partial r}\left(r\frac{\partial v_z}{\partial r}\right)+\frac{\partial^2 v_z}{\partial z^2}\right]. \tag{16.5}$$

In 1921, von Kármán [1] suggested that the partial differential equations could be reduced to ordinary differential equations by seeking a solution of the form

$$v_\theta=rg(z),\ \ v_r=rf(z),\ \ v_z=h(z),\ \ \mathcal{P}=\mathcal{P}(z), \tag{16.6}$$

which is a separation of variables. If these expressions are substituted into Eqs. 16.2 to 16.5, one obtains

$$\left.\begin{aligned}2f+h'&=0,\\ f^2-g^2+hf'&=\nu f'',\\ 2fg+hg'&=\nu g'',\\ \rho hh'+\mathcal{P}'&=\mu h'',\end{aligned}\right\} \tag{16.7}$$

where the primes denote differentiation with respect to z.

The boundary conditions are

$$h = f = 0, \quad g = \Omega \quad \text{at} \quad z = 0.$$
$$f = g = 0 \quad \text{at} \quad z = \infty.$$

The von Kármán transformation is successful in reducing the problem to ordinary differential equations. It should be noted, however, that this solution does not take into account the fact that the radius of the disk might be finite. In practice, these edge effects can frequently be neglected, and the resulting solution is quite useful.

The remaining parameters ν, ρ, and Ω can be eliminated by introducing a dimensionless distance, dimensionless velocities, and a dimensionless pressure as follows:

$$\left.\begin{array}{l} \zeta = z\sqrt{\dfrac{\Omega}{\nu}}, \qquad \mathcal{P} = \mu\Omega P, \\ v_\theta = r\Omega G, \quad v_r = r\Omega F, \quad v_z = \sqrt{\nu\Omega}H. \end{array}\right\} \tag{16.8}$$

The differential Eqs. 16.7 become

$$\left.\begin{array}{r} 2F + H' = 0, \\ F^2 - G^2 + HF' = F'', \\ 2FG + HG' = G'', \\ HH' + P' = H'', \end{array}\right\} \tag{16.9}$$

where the primes now denote differentiation with respect to ζ. The boundary conditions are

$$\left.\begin{array}{l} H = F = 0, \quad G = 1, \quad \text{at} \quad \zeta = 0. \\ F = G = 0 \quad \text{at} \quad \zeta = \infty. \end{array}\right\} \tag{16.10}$$

Since these equations are nonlinear, it seems necessary to obtain the solution numerically [2]. Such a solution to Eqs. 16.9 subject to conditions 16.10 is shown in Fig. 16.1. After the velocity profiles have been determined, the pressure can be obtained by integrating the last of Eqs. 16.9:

$$P = P(0) + H' - \frac{1}{2}H^2. \tag{16.11}$$

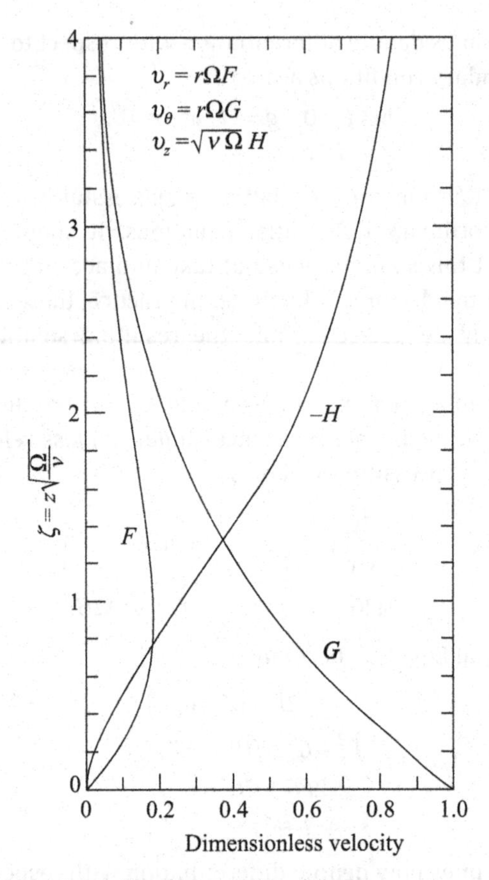

Figure 16.1 Velocity profiles for a rotating disk.

The normal component v_z of the velocity will be important for the calculation of rates of mass transfer to the rotating disk (see Chapter 20). For small distances from the disk, the dimensionless velocity can be expressed as a power series:

$$H = -a\zeta^2 + \frac{1}{3}\zeta^3 + \frac{b}{6}\zeta^4 + \cdots, \tag{16.12}$$

with the coefficients

$$a = 0.51023 \quad \text{and} \quad b = -0.616.$$

On the other hand, for large distances from the disk, the dimensionless velocity can be expressed as

$$H = -\alpha + \frac{2A}{\alpha} e^{-\alpha\zeta} + \cdots, \qquad (16.13)$$

where

$$\alpha = 0.88447 \quad \text{and} \quad A = 0.934.$$

Problems

16.1 A large disk rotates parallel to a stationary plane. The disk and the plane are separated by a distance L [3].

(a) Show that the von Kármán transformation reduces the equations of motion and continuity (for a fluid of constant density and viscosity) to a series of four nonlinear, coupled, ordinary differential equations (thereby permitting an exact solution to the problem for the velocity and pressure distributions) if the dynamic pressure is expressed in the form

$$P = \mu\Omega P(\zeta) + \frac{1}{2}\rho\Omega^2 r^2 Q,$$

where $\zeta = z\sqrt{\Omega / \nu}$, and z is the distance from the disk. Q is an as-yet-undetermined constant. The velocity components retain their forms:

$$v_r = r\Omega F(\zeta), \quad v_\theta = r\Omega G(\zeta), \quad v_z = \sqrt{\nu\,\Omega}\, H(\zeta).$$

(b) Discuss carefully the boundary conditions that need to be satisfied, the order in which the equations are to be solved, and the degrees of freedom that permit the boundary conditions to be satisfied.

(c) How many dimensionless parameters remain in the dimensionless problem? What are they?

16.2 An infinite disk in contact with an infinite expanse of a Newtonian fluid of constant density and viscosity rotates with an angular velocity about an axis perpendicular to the disk. The fluid far from the disk is found to be rotating with an angular velocity ω about the same axis [3].

(a) What equations govern the fluid motion? Be as specific as possible in terms of simplifications which would be invoked to establish a well-defined mathematical problem. However, you need not write out complicated equations in any particular coordinate system.

(b) Analyze the motion far from the disk on the basis of these equations and determine, insofar as possible, expressions for the dynamic pressure $\mathcal{P}$ and the velocity components v_r, v_θ, and v_z in cylindrical coordinates.

(c) Seek an exact solution for the fluid flow pattern and the pressure distribution in the entire flow region. Ignore the possibility of turbulent flow. You may wish to use the von Kármán transformation as a guide. Your work is finished when you have reduced the problem to ordinary differential equations with suitable boundary conditions. Do not attempt to solve such a system further.

References

1. Th. von Kármán. "Über laminare und tuxbulente Reibung." *Zeitschrift fur angewandte Mataematik und Mechanik*, **1**, 233–252 (1921).
2. W. G. Cochran. "The flow due to a rotating disc." *Proceedings of the Cambridge Philosophical Society*, **30**, 365–375 (1934).
3. M. H. Rogers and G. N. Lance. "The rotationally symmetric flow of a viscous fluid in the presence of an infinite rotating disk." *Journal of Fluid Mechanics*, **7**, 617–631 (1960).

Chapter 17

Singular-Perturbation Expansions

Perturbation solutions to complicated problems are sometimes possible when a parameter takes on a very small value. For example, we may be required to solve a differential equation containing such a parameter. If we can solve the equation when the parameter is zero, then we might hope to be able to use this solution as a basis for obtaining correction terms. For a classical perturbation expansion, this is likely to take the form of a power series in the parameter:

$$y(x;\epsilon)=\sum_{n=0}^{\infty}\epsilon^{n}y_{n}(x). \tag{17.1}$$

Frequently, however, the first term $y_0(x)$ is not uniformly valid, that is, there are values of x or regions of x where the first term is grossly in error no matter how small ϵ may be. Such a situation might arise if ϵ multiplied the highest order derivative so that y_0 could not be expected to satisfy all the boundary conditions. This is what happens in boundary-layer flows at high Reynolds number.

A similar failure of the classical perturbation can arise at low Reynolds numbers. The Stokes solution for creeping flow past a sphere is an example. The Stokes solution (see Chapter 15) satisfies the boundary conditions, but examination of the differential equations shows that the inertial terms, which were discarded in obtaining the Stokes solution, become larger than the terms that

The Newman Lectures on Transport Phenomena
John Newman and Vincent Battaglia

ISBN 978-981-4774-27-7 (Hardcover), 978-1-315-10829-2 (eBook)
www.jennystanford.com

were retained (pressure term and viscous term) at sufficiently large distances from the sphere. A power-series expansion such as Eq. 17.1 will not work, as Whitehead pointed out in 1889. The first term, which we calculated in Chapter 15, is all right, but the higher-order terms diverge far from the sphere.

A method of handling these difficulties is to construct two expansions that are valid in different, but overlapping regions of x. These two expansions must match in the region of overlap. In the above example of $y(x;\epsilon)$, we construct one expansion for x near zero and another expansion for large values of x. For this purpose, it is convenient to use different variables in the two regions:

$$\left.\begin{array}{l}\text{inner variable } \bar{x} = \bar{g}(\epsilon)x, \\ \text{outer variable } \tilde{x} = \tilde{g}(\epsilon)x,\end{array}\right\} \tag{17.2}$$

where $\bar{g}(\epsilon)/\tilde{g}(\epsilon) \to \infty$ as $\epsilon \to 0$.

Next, the two expansions can be written formally:

$$\left.\begin{array}{l}\text{inner expansion:} \\ \qquad y = \bar{y}(\bar{x};\epsilon) = \displaystyle\sum_{n=0}^{\infty} \bar{f}_n(\epsilon)\bar{y}_n(\bar{x}), \\ \text{outer expansion:} \\ \qquad y = \tilde{y}(\tilde{x};\epsilon) = \displaystyle\sum_{n=0}^{\infty} \tilde{f}_n(\epsilon)\tilde{y}_n(\tilde{x}),\end{array}\right\} \tag{17.3}$$

where

$$\lim_{\epsilon\to 0} \frac{\bar{f}_{n+1}(\epsilon)}{\bar{f}_n(\epsilon)} = 0 \quad \text{and} \quad \lim_{\epsilon\to 0} \frac{\tilde{f}_{n+1}(\epsilon)}{\tilde{f}_n(\epsilon)} = 0.$$

These expansions 17.3 replace the classical perturbation expansion 17.1. The expansions do not always turn out to be power series in ϵ; therefore, we do not always know in advance the functions $\bar{f}_n(\epsilon)$ and $\tilde{f}_n(\epsilon)$.

The next step in obtaining a perturbation solution would be to substitute the expansions 17.3 (one at a time) into the original statement of the problem for y and to equate the coefficients of equal orders of ϵ. If this involved a differential equation, for example, this step would yield a series of differential equations for the various terms of the expansions. At the same time one would get some information on the nature of the coefficients $\bar{f}_n(\epsilon)$ and $\tilde{f}_n(\epsilon)$.

If the original problem involved a differential equation, then a statement of boundary conditions will be necessary in order to determine the terms of the expansions. The inner expansion $\bar{y}(\bar{x})$ will satisfy any boundary conditions at $x = 0$, while the outer expansion $\tilde{y}(\tilde{x})$ will satisfy any boundary conditions at $x = \infty$. It should be noted that the inner expansion usually does not satisfy the boundary conditions at $x = \infty$, and, conversely, the outer expansion does not usually satisfy the boundary conditions at $x = 0$. Otherwise, we should expect that a classical perturbation would work, and the singular perturbation would not be necessary.

If boundary conditions at $x = 0$ cannot be applied directly to the outer expansion, then the outer expansion cannot be determined completely from differential equations and boundary conditions at $x = \infty$. In such a case, the fact that the inner and outer solutions must match in the region of overlapping validity eliminates any possible ambiguity. This matching condition can be stated as

$$\bar{y}(\bar{x} \to \infty;\epsilon) = \tilde{y}(\tilde{x} \to 0;\epsilon) \tag{17.4}$$

and this condition 17.4 must be satisfied to every order of ϵ. In words, Eq. 17.4 says that the outer limit of the inner expansion is the same as the inner limit of the outer expansion. The meaning will become clearer in the example below. Of course, the inner and outer expansions must satisfy the matching condition 17.4 even if they did not arise from the solution to a differential equation.

Finally, consider the question of a uniformly valid approximation to the correct solution. Suppose that we have the expansions 17.3, which are valid in different, but overlapping regions of x. Then we can form an expansion valid for all x as follows:

$$y(x;\epsilon) = \bar{y}(\bar{x};\epsilon) + \tilde{y}(\tilde{x};\epsilon) - y_c(x;\epsilon) \tag{17.5}$$

where y_c stands for the terms common to both the inner and outer expansions and could be expressed as

$$y_c(x;\epsilon) = \bar{y}(\bar{x} \to \infty;\epsilon) = \tilde{y}(\tilde{x} \to 0;\epsilon). \tag{17.6}$$

It is not difficult to convince oneself that the expansion 17.5, which is called the "composite expansion," is valid for all values of x.

Example

Let us illustrate the application of singular-perturbation expansions by an example drawn from the field of chemical kinetics, namely, the treatment of "active intermediates." This example is selected because of its relative simplicity.

Consider the reaction scheme

$$R \underset{k_1}{\overset{k_o}{\rightleftarrows}} I, \quad I \underset{k_3}{\overset{k_2}{\rightleftarrows}} P.$$

The k's are first-order rate constants that describe the rates of the reactions as follows:

$$\frac{dc_R}{dt} = -k_o c_R + k_1 c_I. \tag{17.7}$$

$$\frac{dc_I}{dt} = -k_2 c_I + k_3 c_P - k_1 c_I + k_o c_R. \tag{17.8}$$

$$\frac{dc_P}{dt} = -k_3 c_P + k_2 c_I. \tag{17.9}$$

Let the initial conditions be

$$c_R = c_R^{\text{o}}, \; c_I = 0, \; c_P = 0 \text{ at } t = 0, \tag{17.10}$$

and let the last rate Eq. 17.9 be replaced by the overall material balance relation

$$c_R + c_I + c_P = \text{const.} = c_R^{\text{o}} \tag{17.11}$$

In this reaction scheme, we regard R as the reactant, I as the intermediate, and P as the product. The case that we consider is one in which I is an "active intermediate." This is true when k_2 is much larger than the other rate constants, so that I, when it is formed, quickly reacts to form P. Under these conditions, the concentration of the intermediate remains small during the entire reaction period. Then one can neglect $k_1 c_I$ in Eq. 17.7 to obtain

$$c_R = c_R^{\text{o}} \, e^{-k_o t}. \tag{17.12}$$

In Eq. 17.8, $k_1 c_I$ and also dc_I/dt should both be small, so that one has approximately

$$c_I = \frac{1}{k_2}(k_o c_R + k_3 c_P). \tag{17.13}$$

The approximation that results in Eqs. 17.12 and 17.13 is known as the quasi-steady-state approximation, and it is quite useful in chemical kinetics. The concentration of the active intermediate, as given by Eq. 17.13, cannot be made to satisfy the initial condition. In order to clarify the mathematical nature of this approximation, Bowen, Acrivos, and Oppenheim [1] applied singular-perturbation methods to the problem and thus provided us with the present example.

In order to eliminate as many parameters as possible, we define the following dimensionless variables and parameters

$$\lambda_1 = k_1/k_o, \quad \lambda_2 = k_2/k_o, \quad \lambda_3 = k_3/k_o, \quad \tau = k_o t$$

$$x = c_R/c_R^o, \quad y = c_I/c_R^o, \quad z = c_P/c_R^o,$$

so that Eqs. 17.11, 17.7, and 17.8 become

$$\left.\begin{aligned} & x+y+z=1, \quad \frac{dx}{d\tau} = -x + \lambda_1 y, \\ & \frac{dy}{d\tau} = \lambda_3 + (1-\lambda_3)x - (\lambda_1+\lambda_2+\lambda_3)y, \end{aligned}\right\} \tag{17.14}$$

and the initial conditions are $x = 1$, $y = 0$ at $\tau = 0$.

To take a specific case, let $\lambda_1 = 1$, $\lambda_3 = 2$, and $\lambda_2 = 1/\epsilon$, where ϵ is the perturbation parameter. When ϵ becomes small, the concentration of I becomes small, and I is an active intermediate. The calculations are summarized in Table 17.1. The inner and outer variables are given, followed by the appropriate differential equations in terms of the inner variable or the outer variable. The inner and outer expansions are written with the functions $\bar{f}_n(\epsilon)$ and $\tilde{f}_n(\epsilon)$ stated explicitly. Usually it is necessary to guess what these functions should be, but in this case simple power series suffice.

Substitution of the expansions into the differential equations yields differential equations for the several terms of the expansions. These can be readily solved. For the outer expansion, one obtains

$$\begin{aligned} \tilde{x} = {} & C_o e^{-\tau} + \epsilon\left[2 + (C_1 - C_o\tau)e^{-\tau}\right] \\ & + \epsilon^2\left[-8 + \left(C_2 - C_1\tau + 2C_o\tau + \frac{1}{2}C_o\tau^2\right)e^{-\tau}\right] + O(\epsilon^3) \end{aligned} \tag{17.15}$$

and

$$\tilde{y} = \epsilon\left(2 - C_o e^{-\tau}\right) + \epsilon^2\left[-8 + (2C_o - C_1 + C_o\tau)e^{-\tau}\right] + O(\epsilon^3). \tag{17.16}$$

The integration constants C_o, C_1, and C_2 are not determined from initial conditions, but by matching with the inner expansion. The inner expansion of x is given in Table 17.1, and its outer limit (as $\bar{\tau} \to \infty$) is

$$x_C = 1 - \epsilon\bar{\tau} + \epsilon^2\left(\frac{1}{2}\bar{\tau}^2 + \bar{\tau} - 1\right) + O(\epsilon^3). \qquad (17.17)$$

The outer expansion of x is given by Eq. 17.15, and its inner limit (as $\tau \to 0$) is

$$\tilde{x}(\tau \to 0) = C_o\left[1 - \tau + \frac{\tau^2}{2} + O(\tau^3)\right]$$
$$+\epsilon[2 + (C_1 - C_o\tau)(1 - \tau + O(\tau^2))]$$
$$+\epsilon^2[-8 + C_2 + O(\tau)] + O(\epsilon^3).$$

Table 17.1 Summary of calculation for example.

	Inner expansion	**Outer expansion**
variable	$\bar{\tau} = \tau/\epsilon.$	$\tilde{\tau} = \tau.$
differential equations	$\frac{dx}{d\bar{\tau}} = -\epsilon x + \epsilon y.$ $\frac{dy}{d\bar{\tau}} = \epsilon(2 - x) - (3\epsilon + 1)y.$	$\frac{dx}{d\tau} = -x + y.$ $\frac{dy}{d\tau} = 2 - x - \left(3 + \frac{1}{\epsilon}\right)y.$
expansions	$\bar{x} = \bar{x}_o(\bar{\tau}) + \epsilon\bar{x}_1(\bar{\tau}) + O(\epsilon^2).$ $\bar{y} = \epsilon\bar{y}_o(\bar{\tau}) + \epsilon^2\bar{y}_1(\bar{\tau}) + O(\epsilon^3).$	$\tilde{x} = \tilde{x}_o(\tau) + \epsilon\tilde{x}_1(\tau) + O(\epsilon^2).$ $\tilde{y} = \epsilon\tilde{y}_o(\tau) + \epsilon^2\tilde{y}_1(\tau) + O(\epsilon^3).$
differential equations for the terms of the expansion	$\frac{d\bar{x}_o}{d\bar{\tau}} = 0, \frac{d\bar{x}_1}{d\bar{\tau}} = -\bar{x}_o,$ $\frac{d\bar{x}_n}{d\bar{\tau}} = -\bar{x}_{n-1} + \bar{y}_{n-2},$ $n = 2, 3, \ldots;$ $\frac{d\bar{y}_o}{d\bar{\tau}} + \bar{y}_o = 2 - \bar{x}_o,$ $\frac{d\bar{y}_n}{d\bar{\tau}} + \bar{y}_n = -\bar{x}_n - 3\bar{y}_{n-1},$ $n = 1, 2, \ldots.$	$\frac{d\tilde{x}_o}{d\tau} + \tilde{x}_o = 0,$ $\frac{d\tilde{x}_n}{d\tau} + \tilde{x}_n = \tilde{y}_{n-1},$ $n = 1, 2, \ldots;$ $\tilde{y}_o = 2 - \tilde{x}_o,$ $\tilde{y}_n = -\tilde{x}_n - 3\tilde{y}_{n-1} - \frac{d\tilde{y}_{n-1}}{d\tau},$ $n = 1, 2, \ldots.$

	Inner expansion	Outer expansion
initial conditions	$\bar{x}_o = 1, \quad \bar{y}_o = 0,$ $\bar{x}_n = 0, \quad \bar{y}_n = 0,$ $n = 1, 2, \ldots.$	There are no initial conditions, but the outer expansion must match the inner expansion.
solutions for the various terms	$\bar{x}_o = 1, \ \bar{x}_1 = -\bar{\tau},$ $\bar{x}_2 = \frac{1}{2}\bar{\tau}^2 + \bar{\tau} + e^{-\bar{\tau}} - 1,$ $\bar{y}_o = 1 - e^{-\bar{\tau}},$ $\bar{y}_1 = \bar{\tau} - 4 + (4 + 3\bar{\tau})e^{-\bar{\tau}}.$	$\tilde{x}_o = e^{-\tau}, \quad \tilde{x}_1 = 2 - (2+\tau)e^{-\tau},$ $\tilde{x}_2 = -8 + \left(\frac{1}{2}\tau^2 + 4\tau + 7\right)e^{-\tau},$ $\tilde{y}_o = 2 - e^{-\tau},$ $\tilde{y}_1 = -8 + (4+\tau)e^{-\tau}.$

In order to put this into a form comparable to Eq. 17.17, we replace τ by $\epsilon\bar{\tau}$ with the result

$$\tilde{x}(\tau \to 0) = C_o + \epsilon(2 + C_1 - C_o\bar{\tau})$$
$$+\epsilon^2\left(-8 + C_2 - C_o\bar{\tau} - C_1\bar{\tau} + \frac{1}{2}C_o\bar{\tau}^2\right) + O(\epsilon^3). \qquad (17.18)$$

Comparison with Eq. 17.17 shows that

$$C_o = 1, \quad C_1 = -2, \quad \text{and} \quad C_2 = 7. \qquad (17.19)$$

These values for the constants have been used in writing the solution in Table 17.1. Note that if we had imposed the initial condition $\tilde{x} = 1$ at $\tau = 0$ on Eq. 17.15, we would have obtained the *incorrect* result $C_2 = 8$.

Finally, we can form a uniformly valid, "composite" expansion by adding the inner and outer expansions and subtracting the common terms:

$$x = \bar{x} + \tilde{x} - x_C$$
$$= e^{-\tau} + \epsilon[2 - (2+\tau)e^{-\tau}] + \epsilon^2\left[e^{-\bar{\tau}} - 8 + \left(\frac{1}{2}\tau^2 + 4\tau + 7\right)e^{-\tau}\right] + O(\epsilon^3)$$
$$(17.20)$$

$$y = \bar{y} + \tilde{y} - y_C$$
$$= \epsilon(2 - \epsilon^{-\tau} - e^{-\bar{\tau}}) + \epsilon^3[-8 + (4+\tau)e^{-\tau} + (4 + 3\bar{\tau})e^{-\tau}] + O(\epsilon^3).$$
$$(17.21)$$

Useful treatments of singular-perturbation expansions can be found in [2, 3].

Problems

17.1 By matching the inner and outer expansions of x, we obtained $C_o = 1$, $C_1 = -2$, $C_2 = 7$ in the example in Chapter 17. Show that for these same values of C_o, C_1, and C_2, the inner and outer expansions of y as given in Table 17.1 and Eq. 17.16 also match.

17.2 Discuss the validity of the outer expansions of the example in Chapter 17 for very long times. If you decide that they are not valid, develop a valid expansion for this region of τ.

17.3 The following differential equation arises in the theory of Debye and Hückel for ionic solutions.

$$\frac{1}{x^2}\frac{d}{dx}\left(x^2\frac{d\psi}{dx}\right) = \epsilon^2 \sinh\left(\psi + \frac{q}{x}\right).$$

$$\frac{d\psi}{dx} = 0 \text{ at } x = 1. \quad \psi \to 0 \text{ as } x \to \infty.$$

ϵ^2 is proportional to the ionic strength, or concentration. x represents radial distance from a central ion, ψ represents the potential due to other ions, and q/x represents the potential of the central ion.

By linearizing the right side,

$$\sinh\left(\psi + \frac{q}{x}\right) \approx \psi + \frac{q}{x},$$

Debye and Hückel obtained the solution

$$\psi = -\frac{q}{x} + \frac{q}{1+\epsilon}\frac{e^{-\epsilon}(x-1)}{x}.$$

The maximum value of $\psi + q/x$ occurs at $x = 1$ and is

$$\psi + q = \frac{q}{1+\epsilon}.$$

The linearization would thus appear to be valid for small q or large ϵ. Nevertheless, the result of Debye and Hückel is generally regarded to be relevant for *small* values of ϵ and to provide the correct limiting law for small concentrations.

The problem for small values of ϵ appears to involve a singular perturbation. Assume and subsequently justify that ψ, the potential due to the ion cloud, rather than $\psi + q/x$ is small for small values of ϵ and develop singular-perturbation expansions for the appropriate inner and outer regions.

To be specific, you are expected to determine the order of magnitude of ψ (in terms of ϵ) in the two regions, to determine the appropriate inner and outer variables $\bar{x} = \bar{g}(\epsilon)x$ and $\tilde{x} = \tilde{g}(\epsilon)x$, and to determine what approximations are valid in each region.

17.4 The Redlich–Kwong equation of state reads

$$P = \frac{RT}{\tilde{V} - b} - \frac{a}{\sqrt{T}\tilde{V}(\tilde{V} + b)}$$

where a and b are constants. For small values of the pressure p, this reduces to the ideal-gas law

$$p\tilde{V} = RT.$$

With this approximate solution as a starting point, develop an expression for the molar volume $\tilde{V}$ as a perturbation expansion for small pressures. Include explicitly terms of order p. (This is the same as including terms of order p^2 in an expansion for the compressibility factor $p\tilde{V}/RT$.) Your procedure should make it clear that you are not neglecting any terms of the desired order.

References

1. J. R. Bowen, A. Acrivos, and A. K. Oppenheim. "Singular perturbation refinement to quasi-steady state approximation in chemical kinetics." *Chemical Engineering Science*, **18**, 177–138 (1963).
2. Andreas Acrivos. "The Method of Matched Asymptotic Expansions." *Chemical Engineering Education*, **2**, 62–65 (1968).
3. Milton Van Dyke. *Perturbation Methods in Fluid Mechanics.* (New York: Academic Press, 1964).

Chapter 18

Creeping Flow past a Sphere

The equations of motion of an incompressible fluid with constant viscosity are given by Eqs. 13.2 and 13.5. A dimensionless statement of the problem of steady, uniform streaming past a body would be

$$\frac{\text{Re}}{2}\mathbf{v}^* \cdot \nabla^* \mathbf{v}^* = -\nabla^* \mathcal{P}^* + \nabla^{*2} \mathbf{v}^*, \tag{18.1}$$

$$\nabla^* \cdot \mathbf{v}^* = 0, \tag{18.2}$$

with boundary conditions

$$\mathbf{v}^* = 0 \text{ on the body,} \tag{18.3}$$

$$\left.\begin{aligned} \mathcal{P}^* &= 0 \\ \mathbf{v}^* &= \mathbf{e}_x \end{aligned}\right\} \text{at infinity,} \tag{18.4}$$

where

$$\mathbf{v}^* = \mathbf{v}/\upsilon_\infty, \quad \mathbf{r}^* = \mathbf{r}/R, \quad \mathcal{P}^* = \frac{\mathcal{P} - \mathcal{P}_\infty}{\mu\upsilon_\infty/R}, \quad \text{Re} = \frac{2R\rho\upsilon_\infty}{\mu}. \tag{18.5}$$

In the problem of uniform streaming, one can vary the shape or orientation of the body and the Reynolds number. Some popular shapes are the sphere and the cylinder along with wedges and other semi-infinite objects and airfoils.

Let us look at the sphere. This problem has never been solved completely and exactly, so attention has been directed at large and

The Newman Lectures on Transport Phenomena
John Newman and Vincent Battaglia

ISBN 978-981-4774-27-7 (Hardcover), 978-1-315-10829-2 (eBook)
www.jennystanford.com

at small values of the parameter Re. The low Reynolds number approximations have been more successful. In 1851, Stokes [1] tried to solve the problem by setting Re = 0 in Eq. 18.1. His solution is given in Eqs. 15.14, 15.15, 15.17, and 15.18.

In 1889, Whitehead tried [2, 3] to extend the Stokes solution to the next term of a classical perturbation expansion. The effort was unsuccessful since the next term cannot be made to satisfy the boundary conditions at infinity. The origin of the difficulty is the fact that the approximation of Stokes is not uniformly valid. Stokes assumed that the inertial terms in Eq. 18.1 were small compared to the pressure term or the viscous terms. However, far from the sphere, the Stokes solution gives (see Problem 15.1)

$$\mathrm{Re}\,\mathbf{v}^* \cdot \nabla^* \mathbf{v}^* = O(\mathrm{Re}/r^{*2}),\ \nabla^* \mathcal{P}^* = O(1/r^{*3}).$$

Thus, no matter how small the Reynolds number may be, the neglected terms are much larger than the retained terms at a sufficiently large value of r^*.

In 1910, Oseen [4, 5] renewed the criticism and suggested an alternative approach. In the equation of motion 18.1, the term on the left is important only when the velocity is nearly equal to the free stream velocity. Thus, he proposed that Eq. 18.1 be modified to read

$$\frac{\mathrm{Re}}{2}\mathbf{e}_x \cdot \nabla^* \mathbf{v}^* = -\nabla^* \mathcal{P}^* + \nabla^{*2} \mathbf{v}^*. \tag{18.6}$$

This has come to be known as the "Oseen equation." The problem thus defined is linear and consequently much simpler than the original problem. Oseen could argue that the term on the left in Eq. 18.6 correctly accounts for the inertial effects in the region where they are important, that is, far from the sphere. On the other hand, near the sphere, the approximation is no worse than the Stokes approximation, which neglects the inertial term altogether. Oseen thus obtained a correction to the Stokes drag:

$$D = 6\pi\mu v_\infty R\left(1 + \frac{3}{16}\mathrm{Re}\right). \tag{18.7}$$

The fact that this correction to the drag is correct to $O(\mathrm{Re})$ should be regarded as fortuitous.

Finally, the problem of low Reynolds number flow past a sphere was treated in 1957 by Proudman and Pearson of Cambridge and independently by Kaplun and Lagerstrom of Cal Tech [6–8]. These authors recognized the singular nature of the perturbation and, therefore, developed two expansions, one valid far from the sphere and one valid near the sphere. These two expansions were then matched in order to supply the missing boundary conditions, and a uniformly valid, composite expansion may be constructed if desired.

For the drag Proudman and Pearson obtained

$$D = 6\pi\mu v_\infty R\left[1 + \frac{3}{16}\mathrm{Re} + \frac{9}{160}\mathrm{Re}^2 \ln \mathrm{Re} + O(\mathrm{Re}^2)\right]. \quad (18.8)$$

The low Reynolds number problem for the circular cylinder is generally considered to be even more difficult than that for the sphere. For the drag per unit length, Kaplun obtained

$$\frac{\text{drag}}{\text{length}} = 4\pi\mu v_\infty \epsilon\left(1 + \sum_{n=2} a_n \epsilon^n\right) + \text{higher order terms}, \quad (18.9)$$

where

$$\epsilon = \left[\frac{1}{2} - \gamma - \ln(\mathrm{Re}/8)\right]^{-1},$$

$\gamma = 0.5772$ is Euler's constant, and $a_2 \approx -0.87$.

Problems

18.1 (a) Stokes law was derived by

- i. Stokes
- ii. Oseen
- iii. Navier
- iv. Poiseuille
- v. Whitehead
- vi. Proudman and Pearson

(b) The Stokes solution for the velocity profiles is subject to the following criticism (indicate all correct answers):

- i. It does not satisfy the boundary condition at $r = \infty$.
- ii. It does not satisfy the boundary condition at $r = R$.
- iii. It does not provide a uniformly valid approximation to the vorticity.
- iv. The terms neglected in the equation of motion are not negligible everywhere in the flow field.
- v. It does not yield the correct drag force for small Reynolds numbers.
- vi. It does not allow corrections to the velocity profiles to be calculated.

vii. The highest order derivatives are dropped in the approximate treatment.

(c) A uniformly valid approximation to the velocity derivatives at low Reynolds numbers was first obtained by

i. Stokes

ii. Oseen

iii. Navier

iv. Poiseuille

v. Whitehead

vi. Proudman and Pearson.

18.2 A manufacturer of large aircraft expresses an interest in "stress corrosion cracking." An idealized sketch of the process is shown below in a coordinate system in which the crack tip is stationary. In this coordinate system, the walls of the crack appear to move with a velocity V, and the process is time independent. Stress corrosion cracking is enhanced in solutions containing chloride ions. Hence, the liquid in the crack can be assumed to be an incompressible, Newtonian liquid of constant viscosity.

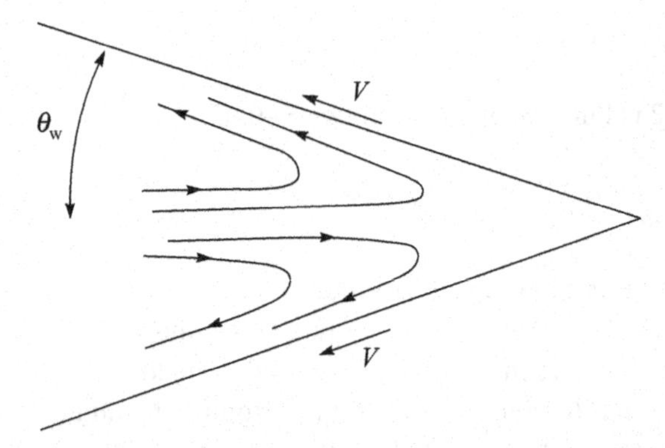

(a) Which coordinate system would appear to be the most convenient for analyzing the velocity and pressure distribution?

i. Rectangular.

ii. Cylindrical.

iii. Spherical.

(b) List all the quantities relevant to the problem and classify each as one of the following:

i. Independent variable.

ii. Dependent variable.

iii. Parameter.

(c) Use the parameters to make the variables dimensionless and classify the resulting quantities as dimensionless independent variables or dimensionless dependent variables.

(d) How many dimensionless parameters remain relevant to the dimensionless problem? What are they?

(e) Which terms in the equation of motion should become negligible very close to the tip of the crack?

i. Inertial. ii. Pressure.

iii. Viscous.

18.3 The physical problem here is the same as that in Problem 18.2, with the emphasis on the region near the tip of the crack.

(a) Assume that the velocity components are, to a first approximation, independent of the distance from the tip of the crack. Use the continuity equation to find the relationship between the two components of the velocity.

(b) What is the order of magnitude of the components of the velocity?

(c) What is the magnitude of the dynamic pressure in the region near the tip of the crack?

(d) Discuss the significance and consequence of the variation of dynamic pressure found in Part (c). (You may prefer to answer this after Part (g).)

(e) Eliminate the dynamic pressure between the two components of the equation of motion. Use the resulting equation to determine the velocity distribution near the crack tip. The results of Part (a) provide the assumed forms for the velocity components.

(f) What boundary conditions should be used to determine the velocity components in Part (e)?

(g) From the equation of motion, now determine explicitly the distribution of dynamic pressure.

References

1. G. G. Stokes. "On the Effect of the Internal Friction of Fluids on the Motion of Pendulums." *Transactions of the Cambridge Philosophical Society*, **9**, 8–106 (1856).

2. A. N. Whitehead. "On the Motion of Viscous Incompressible Fluids." *The Quarterly Journal of Pure and Applied Mathematics*, **23**, 78–93 (1889).

3. A. K. Whitehead. "Second Approximations to Viscous Fluid Motion." *The Quarterly Journal of Pure and Applied Mathematics*, **23**, 143–152 (1889).
4. C. W. Oseen. "Über die Stokes'sche Formel und über eine verwandte Aufgabe in der Hydrodynamik." *Arkiv för Matematik, Astronomi och Fysik*, **6**(29), 1–20 (1911).
5. C. W. Oseen. "Über den Gültigkeitsbereich der Stokesschen Widerstandsformel." *Arkiv för Matematik, Astronomi och Fysik*, **16**, 1–15 (1913).
6. Saul Kaplun and P. A. Lagerstrom. "Asymptotic Expansions of Navier–Stokes Solutions for Small Reynolds Numbers." *Journal of Mathematics and Mechanics*, **6**, 585–593 (1957).
7. Saul Kaplun. "Low Reynolds Number Flow Past a Circular Cylinder." *Journal of Mathematics and Mechanics*, **6**, 595–603 (1957).
8. Ian Proudman and J. R. A. Pearson. "Expansions at small Reynolds numbers for the flow past a sphere and a circular cylinder." *Journal of Fluid Mechanics*, **2**, 237–262 (1957).

Chapter 19

Mass Transfer to a Sphere in Stokes Flow

We want to investigate mass-transfer problems governed by the equation of convective diffusion 13.6. The first example is steady-state mass transfer to a sphere in Stokes flow. By this we mean that the sphere is moving at low Reynolds number so that the velocity distribution in the equation of convective diffusion is given by the Stokes solution 15.14 and 15.15.

For boundary conditions, we choose

$$\left.\begin{aligned} c_i &= c_0 \quad \text{at} \quad r = R. \\ c_i &= c_\infty \quad \text{at} \quad r = \infty. \end{aligned}\right\} \tag{19.1}$$

We choose a constant concentration on the solid surface although other boundary conditions are certainly possible under the appropriate circumstances.

In dimensionless form, the equation of convective diffusion is

$$\frac{\text{Pe}}{2}\mathbf{v}^* \cdot \nabla^*\Theta = \nabla^{*2}\Theta, \tag{19.2}$$

with boundary conditions

$$\Theta = 0 \quad \text{at} \quad r^* = 1, \quad \Theta = 1 \quad \text{at} \quad r^* = \infty, \tag{19.3}$$

where

$$\text{Pe} = 2\upsilon_\infty R/D \quad \text{is the Péclet number.}$$

$$\Theta = \frac{c - c_0}{c_\infty - c_0}, \quad \mathbf{v}^* = \mathbf{v}/\upsilon_\infty, \quad \mathbf{r}^* = \mathbf{r}/R. \tag{19.4}$$

The Newman Lectures on Transport Phenomena
John Newman and Vincent Battaglia

ISBN 978-981-4774-27-7 (Hardcover), 978-1-315-10829-2 (eBook)
www.jennystanford.com

This problem was first treated by Levich [1].

The dimensionless Eq. 19.2 bears a vague resemblance to Eq. 18.1 for the velocity field. For low values of the Péclet number, it is possible to develop a singular-perturbation solution for the concentration field. Acrivos and Taylor [2] give the following asymptotic expression for the average Nusselt number

$$\mathrm{Nu} = 2 + \frac{1}{2}\mathrm{Pe} + \frac{1}{4}\mathrm{Pe}^2 \ln \mathrm{Pe}$$
$$+0.03404\,\mathrm{Pe}^2 + \frac{1}{16}\mathrm{Pe}^3 \ln \mathrm{Pe} + O(\mathrm{Pe}^3), \tag{19.5}$$

where

$$\mathrm{Nu} = \frac{2R\,N_{\mathrm{avg}}}{D(c_\infty - c_0)} = \int_0^\pi \left.\frac{\partial \Theta}{\partial r^*}\right|_{r^*=1} \sin\theta\, d\theta, \tag{19.6}$$

and

N_{avg} is the average flux to the solid (mol/cm^2-s).

Now let us look at large values of Pe. Even though the Reynolds number is less than one, the Péclet number can attain large values since Pe = Re × Sc and Sc = ν/D is about 1000 for liquids. In seeking an approximate solution valid for large Péclet numbers, we are first tempted to neglect diffusion completely in Eq. 19.2, but this is not correct. Convection predominates over diffusion everywhere except in a thin region near the surface of the sphere. Here the derivative of concentration becomes large enough to make the diffusion term comparable to the convective term. This region in which diffusion is important is called the diffusion layer, and an estimate of the order of magnitude of the terms in Eq. 19.2 suggests that its thickness is of order Pe$^{-1/3}$. Consequently, we introduce a stretched coordinate representing distance from the solid surface of the sphere:

$$Y = (r^* - 1)\left(\frac{1}{2}\mathrm{Pe}\right)^{1/3}. \tag{19.7}$$

In spherical coordinates, Eq. 19.2 is

$$\frac{\mathrm{Pe}}{2}\left[\left(1 - \frac{3}{2r^*} + \frac{1}{2r^{*3}}\right)\cos\theta \frac{\partial \Theta}{\partial r^*} - \left(1 - \frac{3}{4r^*} - \frac{1}{4r^{*3}}\right)\frac{\sin\theta}{r^*}\frac{\partial \Theta}{\partial \theta}\right]$$
$$= \frac{1}{r^{*2}}\frac{\partial}{\partial r^*}\left(r^{*2}\frac{\partial \Theta}{\partial r^*}\right) + \frac{1}{r^{*2}\sin\theta}\frac{\partial}{\partial \theta}\left(\sin\theta \frac{\partial \Theta}{\partial \theta}\right).$$

If we replace r^* by Y and let Pe approach infinity, this equation becomes

$$\frac{3}{2}Y^2 \cos\theta \frac{\partial \Theta}{\partial Y} - \frac{3}{2}Y \sin\theta \frac{\partial \Theta}{\partial \theta} = \frac{\partial^2 \Theta}{\partial Y^2} + O(\mathrm{Pe}^{-1/3}). \qquad (19.8)$$

By stretching the normal coordinate according to Eq. 19.7, the convective and diffusion terms have been made to appear of the same order of magnitude in Eq. 19.8.

Let us try to reduce Eq. 19.8 to an ordinary differential equation by means of a similarity transformation:

$$\Theta = \Theta(\eta), \quad \eta = Y/g(\theta), \qquad (19.9)$$

where $g(\theta)$ is to be determined. Substitution of Eq. 19.9 into Eq. 19.8 yields

$$\frac{d^2\Theta}{d\eta^2} - \frac{3}{2}\eta^2(g^3 \cos\theta + g^2 g' \sin\theta)\frac{d\Theta}{d\eta} = 0. \qquad (19.10)$$

One of the requirements for a similarity transformation to work is that the transformed problem should be independent of the original variables, Y and θ, and should involve only the similarity variable η. This can be accomplished for this problem by requiring the coefficient of $\eta^2\Theta'$ in Eq. 19.10 to be constant:

$$g^3 \cos\theta + g^2 g' \sin\theta = \text{const.} = -2. \qquad (19.11)$$

To determine g, multiply Eq. 19.11 by $\sin^2\theta$ and integrate to yield

$$\frac{1}{3}g^3 \sin^3\theta = \frac{1}{2}\sin 2\theta - \theta + \pi, \qquad (19.12)$$

where the integration constant is selected so that g is finite at the upstream end of the boundary layer ($\theta = \pi$).

Equation 19.10 becomes

$$\frac{d^2\Theta}{d\eta^2} + 3\eta^2\frac{d\Theta}{d\eta} = 0, \qquad (19.13)$$

with the boundary conditions

$$\Theta = 0 \text{ at } \eta = 0 \text{ and } \Theta = 1 \text{ at } \eta = \infty, \qquad (19.14)$$

and with the solution

$$\Theta = \frac{1}{\Gamma\left(\frac{4}{3}\right)}\int_0^\eta e^{-x^3}\,dx, \qquad (19.15)$$

where $\Gamma(4/3) = 0.8930$. This solution is displayed in Fig. 19.1.

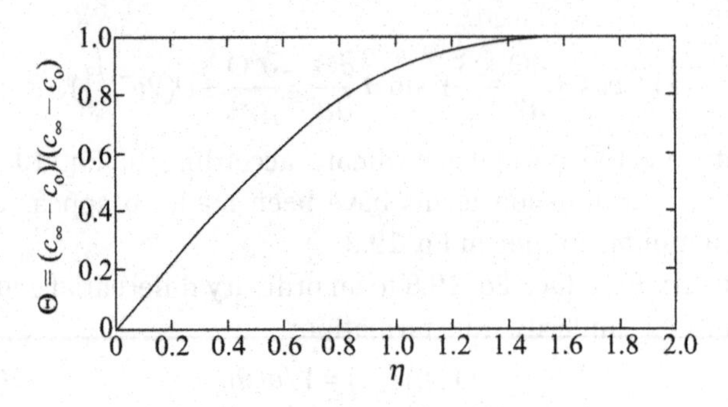

Figure 19.1 Concentration profile in the diffusion layer at high Schmidt numbers.

The dimensionless mass-transfer rate can be determined from

$$\left.\frac{\partial\Theta}{\partial r^*}\right|_{r^*=1} = \frac{\left(\frac{1}{2}\text{Pe}\right)^{1/3}}{g(\theta)}\left.\frac{d\Theta}{d\eta}\right|_{\eta=0} = \frac{\left(\frac{1}{6}\text{Pe}\right)^{1/3}}{\Gamma(4/3)}\frac{\sin\theta}{\left(\pi-\theta+\frac{1}{2}\sin 2\theta\right)^{1/3}}. \tag{19.16}$$

Insertion into Eq. 19.6 thus gives

$$\text{Nu} = \frac{\left(\frac{1}{6}\text{Pe}\right)^{1/3}}{\Gamma(4/3)}\int_0^\pi \frac{\sin^2\theta\, d\theta}{\left(\pi-\theta+\frac{1}{2}\sin 2\theta\right)^{1/3}}. \tag{19.17}$$

Since

$$\int_0^\pi \frac{\sin^2\theta\, d\theta}{\left(\theta-\frac{1}{2}\sin 2\theta\right)^{1/3}} = \frac{1}{2}\frac{3}{2}\left(\theta-\frac{1}{2}\sin 2\theta\right)^{2/3}\Bigg|_0^\pi = \frac{3}{4}\pi^{2/3} = 1.6088,$$

the average Nusselt number becomes

$$\text{Nu} = 0.9914\ \text{Pe}^{1/3}, \tag{19.18}$$

a result due to Levich [1]. Acrivos and Goddard [3] have extended the diffusion-layer solution to higher-order terms and obtain for the average Nusselt number (see Problem 19.2):

$$\text{Nu} = 0.9914\ \text{Pe}^{1/3} + 0.922 + O(\text{Pe}^{-1/3}). \tag{19.19}$$

One should notice that the approximations made in this boundary-layer treatment break down at the rear of the sphere ($\theta = 0$); see Problem 19.1 and Chapter 31.

Problems

19.1 Verify that the diffusion-layer approximations used to treat mass transfer to a sphere in Stokes flow break down at the rear of the sphere ($\theta = 0$) by showing that the diffusion-layer solution does not satisfy the symmetry condition $d\Theta/d\theta = 0$ on the rear axis and by showing that one or more terms neglected in the equation of convective diffusion become important near the rear of the sphere.

19.2 For mass transfer to a sphere in Stokes flow at high Péclet numbers, we found a boundary-layer solution, which was applicable everywhere except at the rear of the sphere, and the mass-transfer rate could be expressed as

$$\text{Nu} = \frac{J/2\pi R}{D(c_\infty - c_0)} = \frac{(4.5\pi^2\text{Pe})^{1/3}}{4\Gamma(4/3)} = 0.9914\,\text{Pe}^{1/3},$$

where $\text{Pe} = 2Rv_\infty/D$ is the Péclet number, J is the total amount of material transferred to the sphere per unit time, c_∞ is the concentration far from the sphere, and c_0 is the (constant) concentration at the surface of the sphere.

Without worrying about the problem at the rear of the sphere, show how one should go about obtaining higher-order corrections to the mass-transfer rate in an expansion valid for large values of Pe by a consideration of the terms neglected in the boundary-layer solution. If you are able to reduce the problem to a set of ordinary differential equations, or if the result is a set of irreducible partial differential equations for the higher-order terms, you are not expected to solve them, but you should indicate the form of the final expression for the mass-transfer rate even though you do not evaluate the constants. Indicate any possible weak points in the development.

19.3 For mass transfer in a spherical coordinate system, the concentration obeys the equation

$$\frac{\partial c_i}{\partial t}+v_r\frac{\partial c_i}{\partial r}+\frac{v_\theta}{r}\frac{\partial c_i}{\partial \theta}+\frac{v_\phi}{r\sin\theta}\frac{\partial c_i}{\partial \phi}$$
$$=\frac{D_i}{r^2}\frac{\partial}{\partial r}\left(r^2\frac{\partial c_i}{\partial r}\right)+\frac{D_i}{r^2\sin\theta}\frac{\partial}{\partial \theta}\left(\sin\theta\frac{\partial c_i}{\partial \theta}\right)+\frac{D_i}{r^2\sin^2\theta}\frac{\partial^2 c_i}{\partial \phi^2}.$$

Formulate a well-posed mathematical problem for determining the steady concentration distribution outside a spherical bubble of radius R rising with a velocity v_∞ in a fluid, where the velocity profile can be taken to be that given at the beginning of Problem 15.2. The concentration at the surface of the bubble is a constant, c_o, corresponding to saturation, and the concentration far from the bubble is c_∞.

Simplify the formulation as much as possible, using dimensionless variables so as to minimize the number of parameters remaining in the problem.

19.4 In Problems 15.2 and 15.3, we considered the irrotational flow past a sphere

$$v_r=\left(1-\frac{R^3}{r^3}\right)v_\infty\cos\theta,\quad v_\theta=-\left(1+\frac{R^3}{2r^3}\right)v_\infty\sin\theta.$$

This exact solution to the Navier–Stokes equation is of some interest in the hydrodynamics of rising bubbles and also in the electrophoretic motion of solid particles and the electrocapillary motion of mercury droplets.

(a) Write out in spherical coordinates the equation of convective diffusion for mass transfer to such a sphere. Assume a steady state and axial symmetry.

(b) Mass transfer to the sphere with $c_i = c_o$ on the surface and $c_i = c_\infty$ far away and with the above velocity distribution will be characterized by several dimensionless parameters. Which one of the following statements most accurately applies to the process?

i. The Reynolds number Re and the Schmidt number Sc must both be given.
ii. The Reynolds number alone is sufficient.
iii. The Schmidt number alone is sufficient.
iv. Only the Péclet number Pe, the product of the Reynolds number and the Schmidt number, need be given.

(c) Is the concept of a *hydrodynamic boundary layer* applicable to the above velocity distribution? (A study of Chapters 21 and 22 may be necessary to answer this part.)

(d) From *each* of the following four pairs, put an asterisk (*) by the member that belongs with "high Pe."

high Pe	low Pe
high Sc	low Sc
high D	low D
Concentration variations occur close to sphere	Concentration variations occur in a region large compared to the sphere

(e) Simplifications are possible within the diffusion layer for small diffusion coefficients. Which of the following is (or are) then incorrect?
i. The effect of curvature of the surface can be ignored in the diffusion term, which then is similar to that for diffusion to a plane.
ii. In the diffusion term, derivatives in the tangential direction can be neglected compared with derivatives in the radial direction.
iii. The convective terms in the equation of convective diffusion can be ignored.
iv. The velocity components can be approximated by their asymptotic forms for large distances from the surface.
v. The problem now becomes parabolic instead of elliptic.

(f) Would you expect the mass-transfer rate to be infinite or finite at the front of the sphere?

(g) Would you expect the mass-transfer rate to be zero or nonzero at the rear of the sphere? Give an answer for the diffusion-layer approximation and for the exact solution to the equation of convective diffusion.

(h) Which one of the following statements would you expect to prevail?

   i. The mass-transfer rate increases monotonically from the front to the rear.
   ii. The mass-transfer rate decreases monotonically from the front to the rear.
   iii. The mass-transfer rate exhibits a maximum somewhere between the front and the rear.
   iv. The mass-transfer rate exhibits a minimum somewhere between the front and the rear.
   v. None of the above applies.

(i) For small diffusion coefficients, how should the normal distance $y = r - R$, from the surface of the sphere, be stretched in order to make the convective and diffusive terms appear to be of the same magnitude within the diffusion layer?

   i. y/D
   ii. $y/D^{1/2}$
   iii. $y/D^{1/3}$
   iv. yD
   v. $yD^{1/2}$
   vi. $yD^{1/3}$

(j) What dependence of flux on diffusion coefficient is implied by this result, and what dependence of Nusselt number on Péclet number is implied?

References

1. Veniamin G. Levich. *Physicochemical Hydrodynamics*. Englewood Cliffs, N. J.: Prentice-Hall, Inc., 1962, Section 14.
2. Andreas Acrivos and Thomas D. Taylor. "Heat and Mass Transfer from Single Spheres in Stokes Flow." *The Physics of Fluids*, **5**, 387–394 (1962).
3. Andreas Acrivos and J. D. Goddard. "Asymptotic expansions for laminar forced convection heat and mass transfer. Part 1. Low speed flows." *Journal of Fluid Mechanics*, **23**, 273–291 (1965).

Chapter 20

Mass Transfer to a Rotating Disk

Now we want to solve the equation of convective diffusion 13.6 for steady-state mass transfer to a rotating disk. The hydrodynamic aspects of the problem were presented in Chapter 16. The pertinent feature of those results is that the velocity normal to the disk, which brings fresh reactant to the surface, depends on z but not on r:

$$v_z = \sqrt{\nu\Omega}\, H\left(\sqrt{\frac{\Omega}{\nu}}z\right). \tag{20.1}$$

Consequently, there is no reason for the concentration to depend on anything besides the normal distance from the disk, and the equation of convective diffusion reduces to

$$v_z \frac{dc_i}{dz} = D\frac{d^2 c_i}{dz^2}, \tag{20.2}$$

with boundary conditions

$$c_i = c_0 \text{ at } z = 0 \quad \text{and} \quad c_i = c_\infty \text{ at } z = \infty. \tag{20.3}$$

Equation 20.2 is a first-order differential equation for dc_i/dz and can be integrated to give

$$\ln\frac{dc_i}{dz} = \frac{1}{D}\int_0^z v_z dz + \text{const}.$$

The Newman Lectures on Transport Phenomena
John Newman and Vincent Battaglia

ISBN 978-981-4774-27-7 (Hardcover), 978-1-315-10829-2 (eBook)
www.jennystanford.com

or

$$\frac{dc_i}{dz} = K \exp\left\{\frac{1}{D}\int_0^z v_z dz\right\}. \tag{20.4}$$

A second integration gives

$$c_i = c_0 + K\int_0^z \exp\left\{\frac{1}{D}\int_0^z v_z dz\right\}dz. \tag{20.5}$$

The constant K is determined from the boundary conditions 20.3:

$$\frac{c_\infty - c_0}{K} = \int_0^\infty \exp\left\{\frac{1}{D}\int_0^z v_z dz\right\}dz = \sqrt{\frac{\nu}{\Omega}}\int_0^\infty \exp\left\{\mathrm{Sc}\int_0^\eta H(\zeta)d\zeta\right\}d\eta.$$

The flux to the disk surface is

$$N = D\left.\frac{dc_i}{dz}\right|_{z=0} = DK$$

or

$$\frac{N}{(c_\infty - c_0)\sqrt{\nu\,\Omega}} = \frac{1}{\mathrm{Sc}\int_0^\infty \exp\left\{\mathrm{Sc}\int_0^\eta H(\zeta)d\zeta\right\}d\eta}. \tag{20.6}$$

The dimensionless mass-transfer rate in Eq. 20.6 is seen to depend only on the Schmidt number Sc = ν/D and is plotted in Fig. 20.1. The calculation is performed by inserting the velocity profile from Chapter 16 into Eq. 20.6.

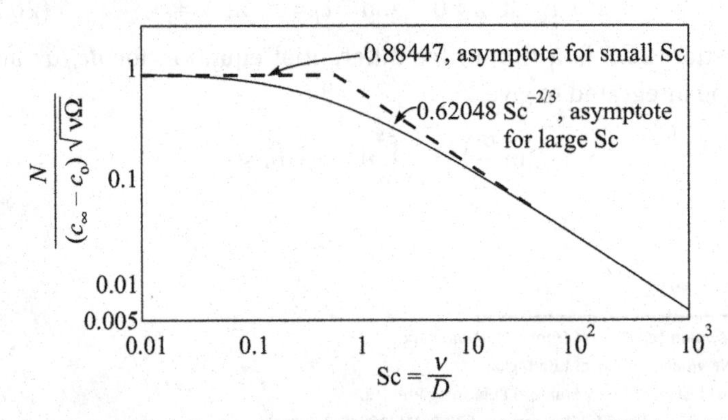

Figure 20.1 Dimensionless mass-transfer rates for a rotating disk.

Levich [1] was the first to treat mass transfer to the rotating disk, even before Wagner [2] treated heat transfer in this system. Sparrow and Gregg [3] provide accurate values for Fig. 20.1.

The asymptotic form for large Schmidt numbers can be obtained by observing that for large Schmidt numbers, the diffusion coefficient D is very small, and the concentration change from c_o to c_∞ takes place very close to the surface of the disk (at small values of ζ in Fig. 16.1 of Chapter 16). Therefore, it is appropriate to use in Eq. 20.6 the velocity profile for small values of ξ, as given in Eq. 16.12. In this way, one obtains [4]

$$\frac{N}{(c_\infty - c_o)\sqrt{\nu\Omega}} = \frac{0.62048\,\mathrm{Sc}^{-2/3}}{1+0.2980\,\mathrm{Sc}^{-1/3}+0.14514\,\mathrm{Sc}^{-2/3}+O(\mathrm{Sc}^{-1})}, \tag{20.7}$$

an expression with a maximum error of about 0.1% in the region Sc > 100.

On the other hand, for very low Schmidt numbers, the diffusion layer extends a large distance from the disk, and it is appropriate to use the velocity profile of Eq. 16.13. At very low Schmidt numbers, Eq. 20.6 becomes

$$\frac{N}{(c_\infty - c_o)\sqrt{\nu\Omega}} = 0.88447e^{-1.611\mathrm{Sc}}\,(1+1.961\,\mathrm{Sc}^2+O(\mathrm{Sc}^3)). \tag{20.8}$$

The first term of Eq. 20.8 tells us that the maximum flux to the disk for very large D is completely determined by the rate of convection of material from infinity:

$$N_{\max} = -(c_\infty - c_o)\sqrt{\nu\Omega}\,H(\infty) = 0.88447\,(c_\infty - c_o)\sqrt{\nu\Omega}. \tag{20.9}$$

Problems

20.1 Derive Eq. 20.7 by means of the velocity profile for distances close to the disk. Note that $\Gamma(4/3)$ = 0.89298 and $\Gamma(5/3)$ = 0.90275.

20.2 Show that the concentration profile in the diffusion layer near a rotating disk is given, at high Schmidt numbers, by Fig. 19.1 when the dimensionless variable η is given by

$$\eta = z(a\nu/3D)^{1/3}\sqrt{\Omega/\nu} = \zeta(a\,\mathrm{Sc}/3)^{1/3}.$$

References

1. B. Levich. "The Theory of Concentration Polarization." *Acta Physicochimica U.R.S.S.*, **17**, 257–307 (1942).
2. Carl Wagner. "Heat Transfer from a Rotating Disk to Ambient Air." *Journal of Applied Physics*, **19**, 837–839 (1948).
3. E. M. Sparrow and J. L. Gregg. "Heat Transfer from a Rotating Disk to Fluids of any Prandtl Number." *Journal of Heat Transfer*, **81C**, 249–251 (1959).
4. John Newman. "Schmidt Number Correction for the Rotating Disk." *The Journal of Physical Chemistry*, **70**, 1327–1328 (1966).

Chapter 21

Boundary-Layer Treatment of a Flat Plate

Let us return to the problem of flow past obstacles mentioned in Chapter 18, but now let us consider flow at high Reynolds numbers, that is, with low viscosity. For steady flow, the equations governing the velocity and pressure distributions are

$$\mathbf{v} \cdot \nabla \mathbf{v} = -\frac{1}{\rho} \nabla \mathcal{P} + \nu \nabla^2 \mathbf{v}, \tag{21.1}$$

$$\nabla \cdot \mathbf{v} = 0. \tag{21.2}$$

If ν is very small, a first approximation to the fluid flow might be obtained by setting $\nu = 0$ in the equation of motion:

$$\mathbf{v} \cdot \nabla \mathbf{v} = -\frac{1}{\rho} \nabla \mathcal{P}. \tag{21.3}$$

This equation is known as the Euler equation and can be solved for the particular problem (or obstacle) at hand. However, since the second-order derivatives of the velocity have been dropped, the boundary condition

$$\mathbf{v} = 0 \text{ on the solid} \tag{21.4}$$

cannot be satisfied. It is possible to make the normal component of the velocity zero:

$$v_n = 0 \text{ on the solid.} \tag{21.5}$$

The Newman Lectures on Transport Phenomena
John Newman and Vincent Battaglia

ISBN 978-981-4774-27-7 (Hardcover), 978-1-315-10829-2 (eBook)
www.jennystanford.com

A solution to the Euler equation satisfying the condition 21.5 is called an Euler solution or an in viscid solution, or an ideal solution. If, in addition, the vorticity is zero, it can be called an irrotational solution or a potential-flow solution. These solutions have the following characteristics:

1. The tangential velocity is, in general, not zero at solid surfaces.
   $$v_t \neq 0 \text{ on the solid.}$$
   This is a reasonable consequence of the assumption of zero viscosity. The fluid can slip past the surface, and a large velocity gradient does not produce any shear stress.
2. The solution is not unique. One can introduce discontinuities of velocity into the flow pattern, or he/she can add a vortex.

An elaborate theory dealing with Euler solutions has been developed. (See, for example, Ref. [1]).

Now we should not attempt to correct the Euler solution to higher-order terms in ν without realizing the singular nature of the perturbation. When ν is small, but not zero, there is a thin layer near the wall in which the velocity changes rapidly from the value predicted by the Euler solution to zero at the wall itself. This region of rapidly changing velocity is called the boundary layer, and in this region, the velocity derivative is so large that $\nu\,\nabla^2\mathbf{v}$ representing the viscous farces is comparable to the pressure forces and the inertial terms in the force balance. Therefore, we can regard the Euler solution as an outer solution, and we must now seek an inner solution valid near the wall.

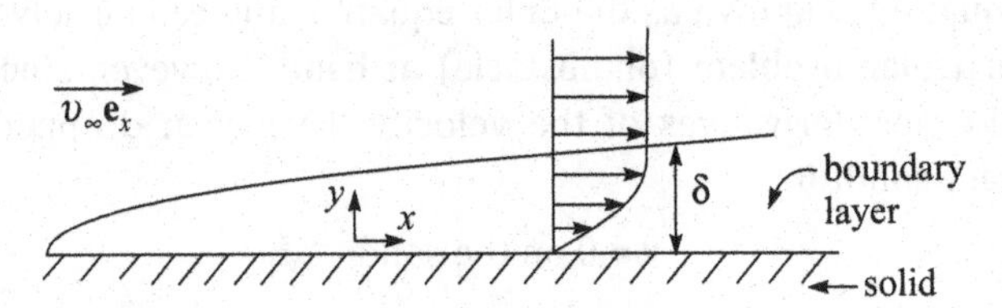

Figure 21.1 Boundary layer on a flat plate.

The flow in the boundary layer is sketched in Fig. 21.1. The boundary-layer thickness δ is a quantity characteristic of the region in which the velocity deviates appreciably from the Euler solution. We want to find how δ depends on ν by comparing terms in the force balance. For concreteness, we look at the boundary layer on a thin, semi-infinite, flat plate, and we let y represent the perpendicular distance from the plate and x represent the distance parallel to the plate, measured from its leading edge.

The continuity equation 21.2 is

$$\frac{\partial \upsilon_x}{\partial x}+\frac{\partial \upsilon_y}{\partial y}=0. \tag{21.6}$$

In the boundary layer, x is of order unity while y is of order δ. In the boundary layer, υ_x varies from zero to the value given by the Euler solution and, hence, is of order unity. Consequently, Eq. 21.6 shows that

$$\upsilon_y = O(\delta)$$

in the boundary layer.

The x-component of the equation of motion 21.1 is

$$\upsilon_x \frac{\partial \upsilon_x}{\partial x}+\upsilon_y \frac{\partial \upsilon_x}{\partial y}=-\frac{1}{\rho}\frac{\partial \mathcal{P}}{\partial x}+\nu\left(\frac{\partial^2 \upsilon_x}{\partial x^2}+\frac{\partial^2 \upsilon_x}{\partial y^2}\right). \tag{21.7}$$

The two terms on the left are of order unity in the boundary layer. The larger of the two viscous terms is

$$\nu\frac{\partial^2 \upsilon_x}{\partial y^2}=O\left(\frac{\nu}{\delta^2}\right).$$

In order for this viscous term to be of the same order as the inertial terms on the left, we must have

$$\delta = O(\nu^{1/2}).$$

The other viscous term is

$$\nu\frac{\partial^2 \upsilon_x}{\partial x^2}=O(\nu)$$

and makes a negligible contribution to the force balance.

The y-component of the equation of motion 21.1 is

$$\upsilon_x \frac{\partial \upsilon_y}{\partial x}+\upsilon_y \frac{\partial \upsilon_y}{\partial y}=-\frac{1}{\rho}\frac{\partial \mathcal{P}}{\partial y}+\nu\left(\frac{\partial^2 \upsilon_y}{\partial x^2}+\frac{\partial^2 \upsilon_y}{\partial y^2}\right). \tag{21.8}$$

The two terms on the left are of order $\delta = 0(\nu^{1/2})$. The viscous terms are

$$\nu\frac{\partial^2 \upsilon_y}{\partial x^2}=O(\nu^{3/2}) \quad \text{and} \quad \nu\frac{\partial^2 \upsilon_y}{\partial y^2}=O(\nu^{1/2}).$$

Hence, Eq. 21.8 shows that

$$\frac{\partial \mathcal{P}}{\partial y} = O(\nu^{1/2})$$

so that the change in $\mathcal{P}$ across the boundary layer is of order ν. Thus, to a first approximation, $\mathcal{P}$ is constant across the boundary layer and depends only on x.

On the basis of these order-of-magnitude arguments, the approximate forms of Eqs. 21.6 and 21.7 appropriate to the boundary layer are

$$v_x \frac{\partial v_x}{\partial x} + v_y \frac{\partial v_x}{\partial y} = -\frac{1}{\rho}\frac{d\mathcal{P}}{dx} + \nu \frac{\partial^2 v_x}{\partial y^2}. \tag{21.9}$$

$$\frac{\partial v_x}{\partial x} + \frac{\partial v_y}{\partial y} = 0. \tag{21.10}$$

These equations determine the two unknowns v_x and v_y. The boundary conditions would be

$$\left.\begin{aligned} v_x = v_y = 0 &\text{ at } y = 0, \\ v_x = U(x) &\text{ at } y = \infty, \end{aligned}\right\} \tag{21.11}$$

where $U(x)$ is the inner limit of the tangential component of the Euler solution. Furthermore, $\mathcal{P}(x)$ is determined by the Euler solution. An additional boundary condition at the upstream end of the boundary layer is required.

After Eqs. 21.9 and 21.10 have been solved for v_x and v_y, a higher approximation to the pressure variation in the boundary layer could be determined from Eq. 21.8.

For a flat plate, the Euler solution is particularly simple:

$$v_x = v_\infty,\ v_y = 0,\ \mathcal{P} = \text{constant}, \tag{21.12}$$

so that

$$U(x) = v_\infty \text{ and } d\mathcal{P}/dx = 0, \tag{21.13}$$

and the boundary-layer problem becomes

$$v_x \frac{\partial v_x}{\partial x} + v_y \frac{\partial v_x}{\partial y} = \nu \frac{\partial^2 v_x}{\partial y^2}, \tag{21.14}$$

$$\frac{\partial v_x}{\partial x} + \frac{\partial v_y}{\partial y} = 0, \tag{21.15}$$

with boundary conditions

$$\left.\begin{aligned} v_x &\to v_\infty \quad \text{as} \quad y \to \infty. \\ v_x &= v_y = 0 \quad \text{at} \quad y = 0. \end{aligned}\right\} \tag{21.16}$$

The problem can be reduced to an ordinary differential equation by means of a similarity transformation:

$$\eta = y\sqrt{\frac{v_\infty}{\nu x}}, \quad v_x = v_\infty f'(\eta), \quad v_y = \frac{1}{2}\sqrt{\frac{\nu v_\infty}{x}}(\eta f' - f), \tag{21.17}$$

where primes denote differentiation with respect to η. By these definitions, the continuity equation is automatically satisfied, and the equation of motion reduces to

$$ff'' + 2f''' = 0 \tag{21.18}$$

with boundary conditions

$$\left.\begin{aligned} f &= 0, \ f' = 0 \quad \text{at} \quad \eta = 0. \\ f' &= 1 \quad \text{at} \quad \eta = \infty. \end{aligned}\right\} \tag{21.19}$$

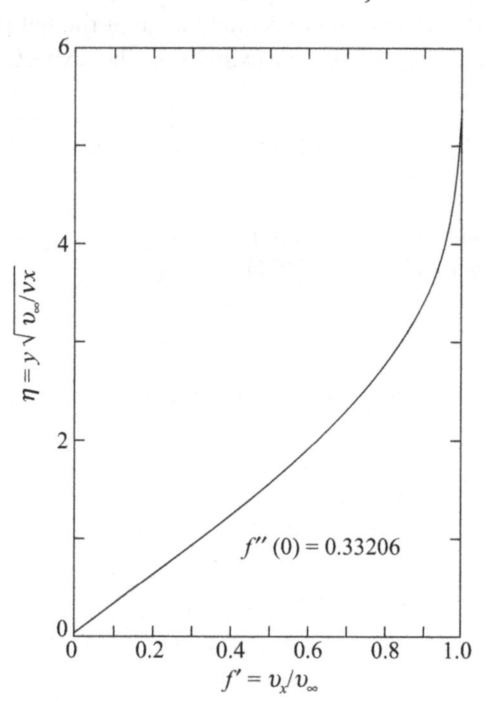

Figure 21.2 Velocity profile for flat plate.

The solution is shown in Fig. 21.2 and confirms our earlier discussion. On the flat plate, there exists a thin layer of thickness

$$\delta \approx 5\sqrt{\frac{\nu x}{\upsilon_\infty}}$$

in which the velocity changes from that predicted from the Euler solution to zero on the surface.

The shear stress at the wall cannot be evaluated from the Euler solution, but the boundary-layer method yields

$$\tau_w = \mu \frac{\partial \upsilon_x}{\partial y}\bigg|_{y=0} = \mu \upsilon_\infty \sqrt{\frac{\upsilon_\infty}{\nu x}} f''(0). \qquad (21.20)$$

This expression agrees with experiment insofar as the flow remains laminar, that is, up to $\upsilon_\infty x/\nu \approx 10^5$.

Problem

21.1 In what region near the leading edge of the flat plate should the boundary-layer approximations be expected to break down?

Reference

1. L. M. Milne-Thomson. *Theoretical Hydrodynamics*, 5th Edition, New York: Dover Publications (2011).

Chapter 22

Boundary-Layer Equations of Fluid Mechanics

In the last chapter we saw that for flow past a flat plate, the equations governing the fluid motion can be simplified if the viscosity has a small value. In this case, viscous effects are important only in a thin region near the solid surface. Similar simplifications are possible for objects of more general shapes.

Consider steady flow past a sphere, and assume that $v_\phi = 0$ and that there are no variations in the ϕ-direction. This is an example of "axisymmetric flow." In spherical coordinates, the equations of motion reduce to:

r-component of the momentum equation

$$v_r \frac{\partial v_r}{\partial r} + \frac{v_\theta}{r}\frac{\partial v_r}{\partial \theta} - \frac{v_\theta^2}{r} = -\frac{1}{\rho}\frac{\partial \mathcal{P}}{\partial r}$$
$$+\nu\left\{\frac{1}{r^2}\frac{\partial}{\partial r}\left(r^2\frac{\partial v_r}{\partial r}\right) + \frac{1}{r^2\sin\theta}\frac{\partial}{\partial \theta}\left(\sin\theta\frac{\partial v_r}{\partial \theta}\right) - \frac{2}{r^2}\left(v_r \frac{\partial v_\theta}{\partial \theta} + v_\theta\cot\theta\right)\right\}. \tag{22.1}$$

The Newman Lectures on Transport Phenomena
John Newman and Vincent Battaglia

ISBN 978-981-4774-27-7 (Hardcover), 978-1-315-10829-2 (eBook)
www.jennystanford.com

θ-component of the momentum equation is

$$v_r \frac{\partial v_\theta}{\partial r} + \frac{v_\theta}{r}\frac{\partial v_\theta}{\partial \theta} + \frac{v_r v_\theta}{r} = -\frac{1}{\rho r}\frac{\partial \mathcal{P}}{\partial \theta}$$

$$+\nu\left\{\frac{1}{r^2}\frac{\partial}{\partial r}\left(r^2\frac{\partial v_\theta}{\partial r}\right) + \frac{1}{r^2 \sin\theta}\frac{\partial}{\partial \theta}\left(\sin\theta\frac{\partial v_\theta}{\partial \theta}\right) + \frac{2}{r^2}\frac{\partial v_r}{\partial \theta} - \frac{v_\theta}{r^2\sin^2\theta}\right\}. \tag{22.2}$$

continuity equation is

$$\frac{1}{r^2}\frac{\partial}{\partial r}(r^2 v_r) + \frac{1}{r\sin\theta}\frac{\partial}{\partial \theta}(v_\theta \sin\theta) = 0. \tag{22.3}$$

We proceed to simplify the problem for $\nu \to 0$ by defining variables appropriate to the boundary layer

$$y = r - R, \quad y' = y/\sqrt{\nu}, \quad v_r' = v_r/\sqrt{\nu}. \tag{22.4}$$

Thus, we assume that the boundary-layer thickness is still of order $\nu^{1/2}$.

If we substitute Eqs. 22.4 into the θ-component of the equation of motion and let ν approach zero, we obtain

$$v_r' \frac{\partial v_\theta}{\partial y'} + \frac{v_\theta}{R}\frac{\partial v_\theta}{\partial \theta} = -\frac{1}{\rho R}\frac{\partial \mathcal{P}}{\partial \theta} + \frac{\partial^2 v_\theta}{\partial y'^2}. \tag{22.5}$$

The r-component of the equation of motion reduces to

$$\frac{1}{\rho}\frac{\partial \mathcal{P}}{\partial y'} = \sqrt{\nu}\frac{v_\theta^2}{R}. \tag{22.6}$$

For a flat plate, $\partial \mathcal{P}/\partial y'$ was of order ν, but for the sphere, curvature effects or centrifugal effects have made $\partial \mathcal{P}/\partial y'$ much larger, of order $\sqrt{\nu}$. Nevertheless, the change in pressure across the boundary layer is still small, and the pressure in the boundary-layer momentum equation 22.5 becomes simply the pressure corresponding to the outer Euler equation.

The continuity equation becomes

$$\frac{\partial v'_r}{\partial y'} + \frac{1}{R\sin\theta}\frac{\partial}{\partial \theta}(v_\theta \sin\theta) = 0. \tag{22.7}$$

A more general axisymmetric body is sketched in Fig. 22.1. The usual discussion of boundary layers on curved objects involves coordinate systems like those shown in the sketch.

- x is the distance measured along the surface from the stagnation point.

- y is the perpendicular distance from the surface.
- $\mathcal{R}(x)$ is the distance of the surface from the axis of symmetry.

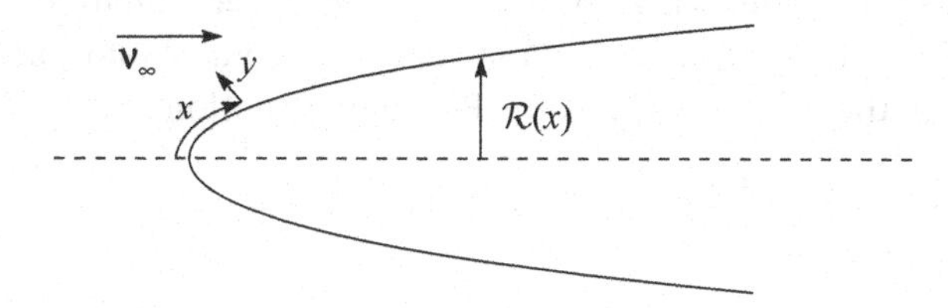

Figure 22.1 Boundary-layer coordinates for an axisymmetric body.

For the sphere discussed above,

$$y = y, \quad x = R(\pi - \theta), \quad \mathcal{R}(x) = R \sin\theta, \quad v_y = v_r, \quad v_x = -v_\theta.$$

Then the above Eqs. 22.5 and 22.7 become

$$v_x \frac{\partial v_x}{\partial x} + v_y \frac{\partial v_x}{\partial y} = -\frac{1}{\rho}\frac{d\mathcal{P}}{dx} + \nu \frac{\partial^2 v_x}{\partial y^2}. \quad (23.8)$$

$$\frac{\partial}{\partial x}(\mathcal{R} v_x) + \frac{\partial}{\partial y}(\mathcal{R} v_y) = 0. \quad (23.9)$$

Axisymmetric boundary-layer equations.

These equations are valid for any axisymmetric boundary-layer flow provided that there are no sharp corners.

For two-dimensional flow near a curved, cylindrical surface, the boundary-layer equations become

$$v_x \frac{\partial v_x}{\partial x} + v_y \frac{\partial v_x}{\partial y} = -\frac{1}{\rho}\frac{d\mathcal{P}}{dx} + \nu \frac{\partial^2 v_x}{\partial y^2}. \quad (23.10)$$

$$\frac{\partial v_x}{\partial x} + \frac{\partial v_y}{\partial y} = 0. \quad (23.11)$$

Two-dimensional boundary-layer equations.

Here y is the perpendicular distance from the surface and x is the distance from the stagnation point measured along the surface.

In both cases, both for axisymmetric flow and for two-dimensional flow, the boundary conditions would be

$$\left.\begin{aligned} v_x = v_y = 0 \ \text{ at } \ y = 0. \\ v_x = U(x) \ \text{ at } \ y = \infty. \end{aligned}\right\} \quad (22.12)$$

The velocity $U(x)$ is the inner limit of the appropriate Euler solution, which describes the fluid motion far from the object.

The momentum equations 22.8 and 22.10 are the same in both cases, and the effect of curvature does not show up explicitly here. The pressure appearing in the boundary-layer equation is the pressure at the solid surface in the appropriate Euler solution and is related to the velocity $U(x)$ by the Bernoulli equation

$$\mathcal{P}+\frac{1}{2}\rho U^2=\text{const.}$$

or

$$\frac{1}{\rho}\frac{d\mathcal{P}}{dx}+U\frac{dU}{dx}=0. \tag{22.13}$$

The continuity equation 22.11 for two-dimensional flow shows no effect of curvature; but the curvature of the surface, represented by $\mathcal{R}(x)$, appears in the continuity equation 22.9 for axisymmetric flow. Another effect of curvature is, as we have said before, to change the pressure difference across the boundary layer from $O(\nu)$ to $O(\sqrt{\nu})$. This applies to both two-dimensional flow and axisymmetric flow.

For two-dimensional flow, the geometry and the external flow affect the boundary layer only through the function $U(x)$. Another simplification results from the fact that the second derivative of v_x with respect to x has been dropped. Thus, the momentum equation becomes parabolic instead of elliptic, so that one can start at the upstream end of the boundary layer and integrate the equations downstream, numerically if necessary. There is no upstream propagation of effects. The boundary-layer equations are still nonlinear, so the problem is not trivial. Nevertheless, a real simplification has been achieved, and considerable attention has been directed toward obtaining solutions of varying degrees of generality for the boundary-layer problem.

Let us summarize the procedure for solving problems of flow past obstacles at high Reynolds numbers. First it is necessary to obtain the appropriate Euler solution by neglecting viscosity altogether. This Euler solution then provides the $U(x)$ and the $\mathcal{P}(x)$, which enter the boundary-layer problem. Within the boundary layer, v_x and v_y are determined from the boundary-layer momentum and continuity equations, Eqs. 22.8 and 22.9 for axisymmetric flows and Eqs. 22.10 and 22.11 for two-dimensional flows. Next it should be possible to obtain a higher-order correction to the external Euler solution, but we shall not discuss this.

Problems

22.1 Show that the following transformation, due to Mangier [1], transforms the problem for an axisymmetric boundary layer into one for a two-dimensional boundary layer:

$$\bar{x} = \frac{1}{L^2}\int_0^x \mathcal{R}^2(x)dx,\ \bar{y} = \frac{\mathcal{R}(x)}{L}y,\ \bar{U} = U,$$

$$\bar{v}_x = v_x,\ \bar{v}_y = \frac{L}{\mathcal{R}}\left(v_y + \frac{\mathcal{R}'(x)}{\mathcal{R}}y\, v_x\right),$$

where L denotes a characteristic length.

22.2 Show under what conditions Bernoulli's equation,

$$\mathcal{P} + \frac{1}{2}\rho v^2 = \text{constant},$$

applies to an inviscid fluid and under what conditions it is valid to replace $d\mathcal{P}/dx$ in the boundary-layer equation by $-\rho U dU/dx$ for the inner limit of the outer, inviscid solution.

22.3 Apply boundary-layer methods to the hydrodynamics of a (convex) cone rotating about its axis with an angular velocity Ω. The situation is similar to that for a rotating disk. However, for the cone, the angle between the axis and the solid surface is somewhat less than $\pi/2$, and here an exact solution to the complete Navier–Stokes equations is not expected.

(a) Is this a two-dimensional or an axisymmetric problem?

(b) What coordinate system would seem to be appropriate for this analysis?

(c) The equation of motion has three components. How will each component be used in formulating the boundary-layer problem?

(d) What parameter should be considered to be large in the analysis? In what region would the analysis be expected to break down?

(e) Formulate differential equations governing the velocity distribution in the boundary layer. How many equations should be solved simultaneously, and for what unknowns?

(f) Formulate boundary conditions appropriate for the differential equations deduced in Part (e).

(g) Reduce the problem to ordinary differential equations by letting

$$v_x = xf(y),\ v_y = h(y),\ \text{and}\ v_z = xg(y),$$

a form inspired by the von Kármán transformation. Here, x and y are boundary-layer coordinates, x being measured along the cone from the apex and y being the perpendicular distance from the surface. The tangential direction perpendicular to x is here labeled z.

(h) Verify that the boundary conditions do not violate the transformation to ordinary differential equations.

(i) Express the solution to the cone problem in terms of the dimensionless functions $F(\zeta)$, $G(\zeta)$, and $H(\zeta)$ in the von Kármán solution for the rotating disk. In particular, show that

$$\beta(x) = \left.\frac{\partial v_x}{\partial y}\right|_{y=0} = x\frac{(\Omega\sin\theta_0)^{3/2}}{\sqrt{\nu}}F'(0),$$

where $F'(0) = a = 0.51023$.

Reference

1. W. Mangler. "Zusammenhang zwischen ebenen und rotationssymmetrichen Grenzschitchen in kompressiblen Fuissigenkeiten." *Zeitschrift fur Angemwandte Mathematik und Mechanik*, **28**, 97–103 (1948).

Chapter 23

Curved Surfaces and Blasius Series

Because of the importance of the boundary-layer problem and its relative simplicity, people have sought "general" solutions to the boundary-layer problem. We shall consider only one illustrative example, the Blasius series for a *symmetric, two-dimensional* flow past a cylindrical surface with a rounded nose.

The object is to obtain generally applicable solutions that can be expressed in terms of tabulated, "universal" functions. The Blasius series involves Taylor series expansions in x about the stagnation point. For a symmetric flow, the external velocity $U(x)$ can be expressed as a power series

$$U(x) = u_1 x + u_3 x^3 + u_5 x^5 + u_7 x^7 + \cdots . \tag{23.1}$$

Because of the assumed symmetry, only the odd powers of x are present. From $U(x)$, the pressure can be obtained according to Eq. 23.13:

$$-\frac{1}{\rho}\frac{d\mathcal{P}}{dx} = u_1^2 x + 4u_1 u_3 x^3 + \cdots . \tag{23.2}$$

Let

$$\eta = y\sqrt{\frac{u_1}{\nu}} \tag{23.3}$$

The Newman Lectures on Transport Phenomena
John Newman and Vincent Battaglia

ISBN 978-981-4774-27-7 (Hardcover), 978-1-315-10829-2 (eBook)
www.jennystanford.com

and

$$\psi = \sqrt{\frac{\nu}{u_1}}\{u_1 x f_1(\eta) + 4u_3 x^3 f_3(\eta) + 6u_5 x^5 f_5(\eta)$$
$$+8u_7 x^7 f_7(\eta) + 10u_9 x^9 f_9(\eta) + 12u_{11} x^{11} f_{11}(\eta) + \cdots\}, \tag{23.4}$$

where ψ is the stream function related to the velocity components by the equations

$$\upsilon_x = \frac{\partial\psi}{\partial y}, \quad \upsilon_y = -\frac{\partial\psi}{\partial x}. \tag{23.5}$$

In order to tabulate functions independent of u_1, u_3, etc., it is necessary to split up the higher-order terms as follows:

$$\left.\begin{aligned}
f_5 &= g_5 + \frac{u_3^2}{u_1 u_5} h_5, \\
f_7 &= g_7 + \frac{u_3 u_5}{u_1 u_7} h_7 + \frac{u_3^3}{u_1^2 u_7} k_7, \\
f_9 &= g_9 + \frac{u_3 u_7}{u_1 u_9} h_9 + \frac{u_5^2}{u_1 u_9} k_9 + \frac{u_3^2 u_5}{u_1^2 u_9} j_9 + \frac{u_3^4}{u_1^3 u_9} q_9, \\
f_{11} &= g_{11} + \frac{u_3 u_9}{u_1 u_{11}} h_{11} + \frac{u_5 u_7}{u_1 u_{11}} k_{11} + \frac{u_3^2 u_7}{u_1^2 u_{11}} j_{11} \\
&\quad + \frac{u_3 u_5^2}{u_1^2 u_{11}} q_{11} + \frac{u_3^3 u_5}{u_1^3 u_{11}} m_{11} + \frac{u_9^5}{u_1^4 u_{11}} n_{11}.
\end{aligned}\right\} \tag{23.6}$$

Thus, the universal functions are f_1, f_3, g_5, h_5, g_7, h_7, k_7, etc. By substituting equations for υ_x and υ_y into the boundary-layer momentum equation 23.10 and equating equal powers of x, we obtain the differential equations for the various universal functions, for example.

$$\left.\begin{aligned}
&f_1''' + f_1 f_1'' - f_1'^2 = -1, \\
&f_3''' + f_1 f_3'' - 4f_1' f_3' + 3f_1'' f_3 = -1, \\
&g_5''' + f_1 g_5'' - 6f_1' g_5' + 5f_1'' g_5 = -1, \\
&h_5''' + f_1 h_5'' - 6f_1' h_5' + 5f_1'' h_5 = -\frac{1}{2} + 8(f_3'^2 - f_3 f_3'').
\end{aligned}\right\} \tag{23.7}$$

The boundary conditions are

$$\left.\begin{aligned} &0 = f_1 = f_1' = f_3 = f_3' = g_5 = g_5' = h_5 = h_5', \text{etc., at } \eta = 0. \\ &f_1' = 1, f_3' = 1/4, g_5' = 1/6, h_5' = 0, \text{etc.,} \quad \text{at } \eta = \infty. \end{aligned}\right\} \quad (23.8)$$

The solutions to Eqs. 23.7 and 23.8 for the universal functions have been calculated and are tabulated in Table 23.1. These tables are essentially a duplication of those calculated by Tifford (WADC-TR-53-288, Part 4.). See also Ref. [1].

Table 23.1 Universal functions of Blasius series.

η	f_1'	f_3'	g_5'	h_5'	g_7'	h_7'	k_7'
0.00	0.00	0. 00	0. 00	0. 00	0. 00	0. 00	0. 00
0.20	0.22661	0.12510	0.10722	0.01413	0.09617	0.01732	0.00164
0.40	0.41446	0.21288	0.17781	0.01170	0.18632	0.00298	0.00439
0.60	0.56628	0.26881	0.21837	-0.00106	0.18793	-0.02864	0.00966
0.80	0.68594	0.29972	0.23665	-0.01767	0.19935	-0.06369	0.01743
1.00	0.77787	0.31250	0.23990	-0.03306	0.19798	-0.09248	0.02709
1.20	0.84667	0.31327	0.23415	-0.04419	0.18958	-0.11019	0.03691
1.40	0.89681	0.30702	0.22390	-0.04989	0.17820	-0.11594	0.04518
1.60	0.93235	0.29746	0.21226	-0.05040	0.166545	-0.11144	0.05060
1.80	0.95683	0.28712	0.20115	-0.04682	0.15580	-0.09968	0.05251
2.00	0.97322	0.27755	0.19158	-0.04061	0.14694	-0.08388	0.05097
2.20	0.98385	0.26950	0.18392	0.03317	0.14001	-0.06688	0.04657
2.40	0.99055	0.26322	0.17814	-0.02568	0.13489	-0.05070	0.04024
2.60	0.99463	0.25859	0.17401	-0.01891	0.13128	-0.03669	0.03297
2.80	0.99705	0.25596	0.17119	-0.1327	0.12884	-0.02538	0.02569
3.00	0.99842	0.25322	0.16935	-0.00890	0.12726	-0.01681	0.01906
3.20	0.99919	0.25186	0.16820	-0.00571	0.12629	-0.01067	0.01349
3.40	0.99959	0.25103	0.16751	-0.00351	0.12571	-0.00649	0.00911
3.60	0.99980	0.25055	0.16711	-0.00206	0.12537	-0.00379	0.00588
3.80	0.99991	0.25028	0.16690	-0.00116	0.12519	-0.00212	0.00362
4.00	0.99996	0.25014	0.16678	-0.00063	0.12509	-0.00114	0.00214
4.20	0.99998	0.25007	0.16672	-0.00033	0.12504	-0.00059	0.00120
4.40	0.99999	0.25003	0.16669	-0.00016	0.12502	-0.00029	0.00065

η	g_9'	h_9'	k_9'	j_9'	q_9'
0. 00	0. 00	0. 00	0. 00	0. 00	0. 00
0.20	0.08837	0.01125	0.00188	0.01253	-0.00616
0.40	0.14125	-0.00792	-0.01123	0.02878	-0.01239
0.60	0.16686	-0.04173	-0.03105	00.5246	-0.01895
0.80	0.17398	-0.07602	-0.05014	0.08328	-0.02617
1.00	0.17002	-0.10187	-0.06386	0.11706	-0.03409
1.20	0.16044	-0.11566	-0.07044	0.14804	-0.04235
1.40	0.14888	-0.11760	-0.07027	0.17104	-0.05018
1.60	0.13755	-0.11013	-0.06486	0.18286	-0.05668
1.80	0.12758	-0.09648	-0.05616	0.18269	-0.06097
2.00	0.11946	-0.07983	-0.04600	0.17178	-0.06246
2.20	0.11322	-0.06275	-0.03585	0.15279	-0.06098
2.40	0.10866	-0.04704	-0.02667	0.12902	-0.05675
2.60	0.10547	-0.03371	-0.01898	0.10369	-0.05036
2.80	0.10333	-0.02313	-0.01295	0.07946	-0.04263
3.00	0.10196	-0.01522	-0.00847	0.05813	-0.03442
3.20	0.10111	-0.00960	-0.00532	0.04063	-0.02652
3.40	0.10061	-0.00581	-0.00321	0.02715	-0.01950
3.60	0.10032	-0.00338	-0.00186	0.01735	-0.01368
3.80	0.10016	-0.00189	-0.00103	0.01061	-0.00916
4.00	0.10008	-0.00101	-0.00055	0.00621	-0.00586
4.20	0.10004	-0.00052	-0.00028	0.00348	-0.00358
4.40	0.10002	-0.00026	-0.00014	0.00187	-0.00209

η	g_{11}'	h_{11}'	k_{11}'	j_{11}'	q_{11}'	m_{11}'	n_{11}'
0.00	0. 00	0. 00	0. 00	0. 00	0. 00	0. 00	0. 00
0.20	0.08243	0.00743	-0.00413	0.01654	0.02371	-0.03593	0.01031
0.40	0.12985	-0.01448	-0.03681	0.03710	0.05140	-0.07210	0.02056
0.60	0.15111	-0.04887	-0.07989	0.6508	0.08632	-0.10947	0.03069
0.80	0.15531	-0.08159	-0.11805	0.09917	0.12667	-0.14885	0.04064
1.00	0.14979	-0.10456	-0.14281	0.13419	0.16663	-0.18951	0.05030
1.20	0.13970	-0.11522	-0.15196	0.16408	0.19951	-0.22877	0.05947

η	g'_{11}	h'_{11}	k'_{11}	j'_{11}	q'_{11}	m'_{11}	n'_{11}
1.40	0.12833	-0.11459	-0.14741	0.18408	0.22023	-0.26272	0.06776
1.60	0.11754	-0.10547	-0.13304	0.19187	0.22656	-0.28735	0.07466
1.80	0.10827	-0.09114	-0.11307	0.18756	0.21899	-0.29982	0.07963
2.00	0.10082	-0.07457	-0.09120	0.17310	0.20013	-0.29815	0.08218
2.20	0.09515	-0.05809	-0.07014	0.15154	0.17368	-0.28340	0.08199
2.40	0.09104	-0.04322	-0.05161	0.12626	0.14356	-0.25748	0.07898
2.60	0.08818	-0.03078	0.03639	0.10032	0.11326	-0.22368	0.07334
2.80	0.08628	-0.02102	-0.02462	0.07613	0.08546	-0.18583	0.06553
3.00	0.08506	-0.01376	-0.01600	0.05524	0.06101	-0.14763	0.05625
3.20	0.08431	-0.00865	-0.00999	0.03834	0.04254	-0.11214	0.04631
3.40	0.08387	-0.00522	-0.00599	00.2546	0.02813	-0.08144	0.03651
3.60	0.08361	-0.00303	-0.00345	0.01619	0.01782	-0.05654	0.02754
3.80	0.8348	-0.00168	-0.00191	0.00986	0.01081	-0.03753	0.01986
4.00	0.08340	-0.00090	-0.00102	0.00575	0.00628	-0.02380	0.01368
4.20	0.08337	-0.00046	-0.00052	0.00321	0.00350	-0.01443	0.00899
4.40	0.08335	-0.00023	-0.00026	0.00172	0.00187	-0.00837	0.00564

$f''_1(0)$	$f''_3(0)$	$g''_5(0)$	$h''_5(0)$	$g''_7(0)$	$h''_7(0)$	$k''_7(0)$
1.23259	0.72445	0.63470	0.11919	0.57920	0.18296	0.00764

$g''_9(0)$	$h''_9(0)$	$k''_9(0)$	$j''_9(0)$	$q''_9(0)$
0.53993	0.15198	0.05719	0.06074	-0.03079

$g''_{11}(0)$	$h''_{11}(0)$	$k''_{11}(0)$	$j''_{11}(0)$	$q''_{11}(0)$	$m''_{11}(0)$	$n''_{11}(0)$
0.50999	0.13230	0.07421	0.08055	0.11636	-0.17964	0.05156

Reference

1. Hermann Schlichting, *Boundary-Layer Theory*, translated by J. Kestin. New York: McGraw-Hill Book Company, 1979, pp. 168–173.

Chapter 24

The Diffusion Boundary Layer

The preceding three chapters deal with the calculation of velocity profiles in boundary layers, that is, near surfaces at high Reynolds numbers. These results are useful to chemical engineers particularly for the calculation of mass-transfer rates. The rate of mass transfer depends, of course, on the convective flow pattern near the solid surface, especially at high Schmidt numbers.

The concentration distribution is determined by the equation of convective diffusion 13.6. In terms of dimensionless velocity and coordinates, the steady-state form of this equation is

$$\mathbf{v}^* \cdot \nabla^* c_i = \frac{1}{\mathrm{Pe}} \nabla^{*2} c_i , \tag{24.1}$$

where

$$\mathbf{v}^* = \mathbf{v}/v_\infty, \quad \mathbf{r}^* = \mathbf{r}/l,$$

l is a length characteristic of the body, and $\mathrm{Pe} = v_\infty l/\mathcal{D}$ is the Péclet number. Since the Péclet number is equal to the Reynolds number times the Schmidt number,

$$\mathrm{Pe} = \mathrm{Re}\,\mathrm{Sc} = \frac{v_\infty l}{\nu}\frac{\nu}{\mathcal{D}}, \tag{24.2}$$

a large value of the Reynolds number implies a large value for the Péclet number unless the Schmidt number is extremely small.

The Newman Lectures on Transport Phenomena
John Newman and Vincent Battaglia

ISBN 978-981-4774-27-7 (Hardcover), 978-1-315-10829-2 (eBook)
www.jennystanford.com

In Chapter 19, we have already considered diffusion to a sphere in Stokes flow when the Péclet number is large. Now we consider diffusion to solids in boundary-layer flows, that is, at high Reynolds numbers. The present problem is similar to the earlier one. Convective effects are dominant over diffusion almost everywhere, but near the solid surface, diffusion must also become important. The result is a mass-transfer boundary layer not very different from that encountered on the sphere in Stokes flow. However, here we consider the Péclet number to be large because the Reynolds number is large. The velocity profiles are then those given by boundary-layer theory.

As long as the Péclet number is large, the equation of convective diffusion reduces to

$$v_x^* \frac{\partial c_i}{\partial x^*} + v_y^* \frac{\partial c_i}{\partial y^*} = \frac{1}{\text{ReSc}} \frac{\partial^2 c_i}{\partial y^{*2}}. \tag{24.3}$$

This is the boundary-layer equation applicable near the surface of the solid, and x^* and y^* are the usual boundary-layer coordinates defined in Chapter 22, but made dimensionless with l. Far from the surface, diffusion is negligible, and the outer solution is $c_i = c_\infty$.

Equation 24.3 contains some simplifications similar to those in the boundary-layer equations for fluid mechanics. An elliptic equation has become parabolic, with the consequence that there is no upstream propagation of effects and no need for a downstream boundary condition. This also means that the equation could be integrated numerically by stepping downstream from the beginning of the mass-transfer section. Also similar to the tangential momentum equation is the fact that curvature effects do not show up explicitly and one form of the equation applies to all problems. The problem remains linear once the velocity profiles are given, and the mass-transfer rate is influenced by the same things that influence the velocity profiles, namely, the Euler velocity $U(x)$ adjacent to the surface and (for axisymmetric problems) the distance of the surface from the axis of symmetry, as well as the Schmidt number and the boundary conditions for the mass-transfer part of the problem.

For the boundary condition on the solid surface, we could have a specified concentration, a specified flux, or a specified relationship between the flux and the concentration. We shall consider only the case $c_i = c_0$, a constant, on the surface. (Chapter 29 treats reaction kinetics under certain conditions for a fluid flowing past a catalytic surface in the form of a flat plate, having the hydrodynamics described in Chapter 21.)

The explicit appearance of the Reynolds number or of the parameter ν in the problem will disappear if we stretch the normal coordinate and the normal component of the velocity as follows:

$$y' = y^* \mathrm{Re}^{1/2}, \quad v_y' = v_y^* \mathrm{Re}^{1/2}. \tag{24.4}$$

(This is in accord with the magnitudes of these variables for boundary-layer flows, as determined in Chapter 22.) Then the boundary-layer equations become

$$v_x^* \frac{\partial v_x^*}{\partial x^*} + v_y' \frac{\partial v_x^*}{\partial y'} = U^* \frac{dU^*}{dx^*} + \frac{\partial^2 v_x^*}{\partial y'^2}, \tag{24.5}$$

$$\frac{\partial v_x^*}{\partial x^*} + \frac{\partial v_y'}{\partial y'} = 0 \text{ (two-dimensional)}, \tag{24.6}$$

$$v_x^* \frac{\partial c_i}{\partial x^*} + v_y' \frac{\partial c_i}{\partial y'} = \frac{1}{\mathrm{Sc}} \frac{\partial^2 c_i}{\partial y'^2}. \tag{24.7}$$

The convective velocities v_x^* and v_y' in Eq. 24.7 are now functions of x^* and y' and do not depend explicitly on the Reynolds number (or on ν).

We still have the Schmidt number as a parameter. For Sc = 1, we expect the thickness of the diffusion layer to be about the same as the thickness of the hydrodynamic boundary layer. For large Sc, the diffusion coefficient is small, and the diffusion layer becomes much thinner than the hydrodynamic boundary layer. Then only the boundary-layer flow very near the surface is important, and the velocity components can be approximated as

$$v_x^* = B(x^*)y' \quad \text{and} \quad v_y' = -\frac{1}{2}\frac{dB}{dx^*} y'^2, \tag{24.8}$$

where

$$B(x^*) = \left. \frac{dv_x^*}{\partial y'} \right|_{y'=0}. \tag{24.9}$$

These are the first terms in power-series expansions of the velocity components in the distance from the wall; the coefficient of v_y' follows from the continuity equation 24.6 or 22.11 for two-dimensional boundary layers. Equation 24.7 can be written as

$$B(x^*)y'\frac{\partial c_i}{\partial x^*}-\frac{1}{2}\frac{dB}{dx^*}y'^2\frac{\partial c_i}{\partial y'}=\frac{1}{\text{Sc}}\frac{\partial^2 c_i}{\partial y'^2}. \tag{24.10}$$

If, for large Schmidt numbers, we make the substitution

$$Y=y'\,\text{Sc}^{1/3}, \tag{24.11}$$

then the mass-transfer, boundary-layer equation becomes

$$B(x^*)Y\frac{\partial c_i}{\partial x^*}-\frac{1}{2}\frac{dB}{dx^*}Y^2\frac{\partial c_i}{\partial Y}=\frac{\partial^2 c_i}{\partial Y^2}. \tag{24.12}$$

Then c_i is a function only of x^* and Y and is nominally independent of Sc, as well as of Re. Consequently, the dependence of the mass-transfer rate on the Reynolds and Schmidt numbers can be given explicitly by:

$$\mathcal{D}\frac{\partial c_i}{\partial y}=\frac{\mathcal{D}}{l}\frac{\partial c_i}{\partial y^*}=\frac{\mathcal{D}}{l}\text{Re}^{1/2}\frac{\partial c_i}{\partial y'}=\frac{\mathcal{D}}{l}\text{Re}^{1/2}\text{Sc}^{1/3}\frac{\partial c_i}{\partial Y}. \tag{24.13}$$

On the other hand, when $\text{Sc}\to 0$, the diffusion boundary layer is much thicker than the hydrodynamic boundary layer, and then the external inviscid flow determines the important part of the convective velocity in the diffusion layer. This situation would apply to heat transfer in liquid metals where the Prandtl number (analogous to the Schmidt number in mass transfer) can be on the order of 0.01. Note that it is still necessary for the Péclet number to be large in order for the boundary-layer form of the equation of convective diffusion 24.3 to be applicable. The Nusselt number is then proportional to the square root of the Schmidt number (see Problem 24.3).

The dependence of the Nusselt number (a dimensionless mass-transfer rate, $\text{Nu} = (l/\mathcal{D})\partial c_i/\partial y|_{y=0}$) on the Schmidt number is sketched in Fig. 24.1, which shows the behavior for both high and low Sc.

The Lighthill transformation. It has been suggested [1, 2] that the exact solution to this problem at high Schmidt numbers can be obtained when the concentration is constant over the surface, that is, when the boundary conditions are

$$c_i = c_0 \text{ at } y=0 \quad\text{and}\quad c_i\to c_\infty \text{ as } y\to\infty. \tag{24.14}$$

For greater generality, let us rewrite the problem in dimensional form after it had been reduced to Eq. 24.12. Thus, we start with the diffusion-layer form appropriate for two-dimensional problems:

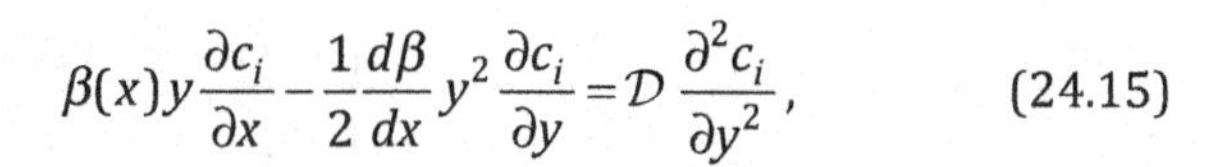

$$\beta(x)y\frac{\partial c_i}{\partial x} - \frac{1}{2}\frac{d\beta}{dx}y^2\frac{\partial c_i}{\partial y} = \mathcal{D}\frac{\partial^2 c_i}{\partial y^2}, \tag{24.15}$$

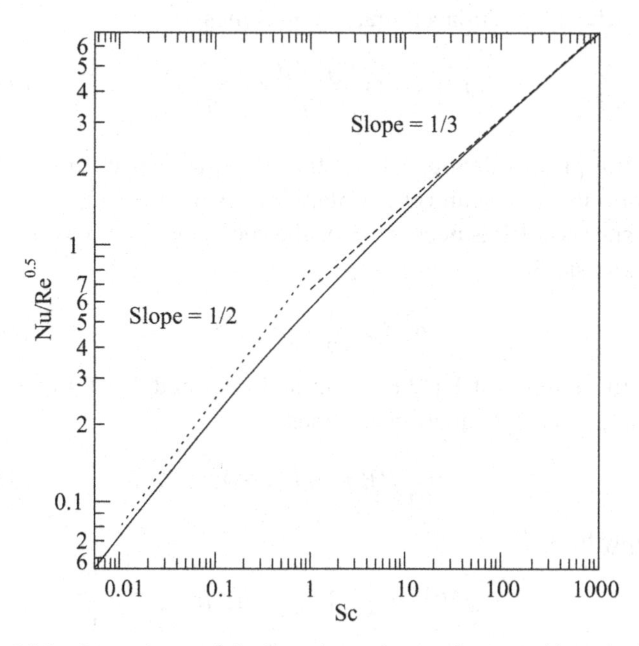

Figure 24.1 Dependence of the Nusselt number on the Reynolds and Schmidt numbers for boundary-layer flows, showing the slopes of the asymptotes at high and low Schmidt numbers. The numbers are taken from the second column of Table 26.1 and the first entries of Tables 26.2 and 26.3.

where

$$\beta(x) = \left.\frac{\partial v_x}{\partial y}\right|_{y=0}. \tag{24.16}$$

Seek a similarity solution of the form

$$\frac{c_i - c_0}{c_\infty - c_0} = \Theta(\eta) \text{ where } \eta = y/g(x). \tag{24.17}$$

Then

$$\frac{\partial c_i}{\partial x} = -(c_\infty - c_0)\frac{d\Theta}{d\eta}\frac{g'(x)}{g^2(x)}y, \tag{24.18}$$

$$\frac{\partial c_i}{\partial y} = (c_\infty - c_0)\frac{d\Theta}{d\eta}\frac{\partial \eta}{\partial y} = \frac{c_\infty - c_0}{g(x)}\frac{d\Theta}{d\eta}, \tag{24.19}$$

$$\frac{\partial^2 c_i}{\partial y^2} = \frac{c_\infty - c_0}{g^2(x)} \frac{d^2\Theta}{d\eta^2}, \tag{24.20}$$

so that the diffusion-layer equation becomes

$$\Theta'' = -\eta^2\Theta' \left[\frac{g^2 g' \beta}{\mathcal{D}} + \frac{1}{2} \frac{g^3 \beta'}{\mathcal{D}} \right]. \tag{24.21}$$

Here the primes denote differentiation with respect to x or η, as appropriate, for g, β, and Θ. In order for the similarity transformation to be successful, it is necessary for the coefficient in brackets to be a constant, say 3:

$$g^2 g' \beta + \frac{1}{2} g^3 \beta' = 3\mathcal{D}. \tag{24.22}$$

It turns out that Eq. 24.22 can be integrated. Let $f = g^3$ so that $f' = 3g^2 g'$, and this equation becomes

$$\frac{1}{3} f' \beta + \frac{1}{2} f \beta' = 3\mathcal{D}. \tag{24.23}$$

Multiply by $3\beta^{1/2}$,

$$f' \beta^{3/2} + \frac{3}{2} f \beta^{1/2} \beta' = 9\mathcal{D} \beta^{1/2}, \tag{24.24}$$

and integrate

$$f \beta^{3/2} = 9\mathcal{D} \int_0^x \beta^{1/2} dx. \tag{24.25}$$

Thus

$$g = \frac{(9\mathcal{D})^{1/3}}{\beta^{1/2}} \left[\int_0^x \beta^{1/2} dx \right]^{1/3} \tag{24.26}$$

and

$$\eta = \frac{y\beta^{1/2}}{\left[9\mathcal{D} \int_0^x \beta^{1/2} dx \right]^{1/3}}. \tag{24.27}$$

The integration constant in Eq. 24.25 can be set to zero for both bluff and pointed bodies, although for different reasons. For a pointed body (for example, a fiat plate), the diffusion layer starts at zero thickness (meaning that f is zero at $x = 0$). For bluff bodies, like a circular cylinder, β is proportional to x near $x = 0$. A diffusion layer beginning in an already developed boundary layer also has a zero value of f at the beginning of the mass-transfer section. In general, the selection of the integration constant is related to the success of the similarity transformation in satisfying all the boundary conditions of the original problem embodied in Eq. 24.15.

Now the differential equation for Θ becomes

$$\Theta'' + 3\eta^2\Theta' = 0 \tag{24.28}$$

with the solution

$$\Theta = \frac{1}{\Gamma(4/3)}\int_0^{\eta} e^{-\eta^3}\,d\eta. \tag{24.29}$$

Hence

$$\left.\frac{\partial c_i}{\partial y}\right|_{y=0} = \frac{(c_\infty - c_0)\beta^{1/2}/l}{\left[9\mathcal{D}\int_0^x \beta^{1/2}dx\right]^{1/3}}\left.\frac{d\Theta}{d\eta}\right|_{\eta=0} = \frac{(c_\infty - c_0)\beta^{1/2}/l}{\Gamma(4/3)\left[9\mathcal{D}\int_0^x \beta^{1/2}dx\right]^{1/3}}, \tag{24.30}$$

where l is a characteristic length. The average Nusselt number over a distance from $x = 0$ to $x = L$ is

$$\mathrm{Nu} = \frac{1}{L}\int_0^L \mathrm{Nu}_x dx = \frac{3l}{2L}\frac{\left[\int_0^L \beta^{1/2}dx\right]^{2/3}}{\Gamma(4/3)(9\mathcal{D})^{1/3}}. \tag{24.31}$$

The solution obtained is really quite remarkable since an exact solution for large Sc has been obtained for an arbitrary variation of β with distance x. Note that β is simply related to the wall shear stress in the direction of flow, a quantity that is likely to be reported as an important result of an investigation of the fluid mechanics of a problem. We can also see that β cannot become negative and still give a meaningful result. Such a value of β would be obtained for values of x beyond the point of separation, where the boundary-layer approach is no longer valid, either for the fluid mechanics or for the mass transfer.

Problems

24.1 Ascertain the asymptotic form for large Schmidt number of the local Nusselt number at the forward stagnation point of a circular cylinder. Assume that the inner limit of the Euler solution is

$$U = 2v_\infty \sin\theta$$

(which corresponds to the continuous, irrotational Euler solution). This is the tangential velocity outside the hydrodynamic boundary layer. The local Nusselt number is defined as

$$\mathrm{Nu} = \frac{2R}{D}\cdot\frac{\text{flux}}{c_\infty - c_0} = \frac{2R}{c_\infty - c_0}\left.\frac{\partial c_i}{\partial r}\right|_{r=R}.$$

24.2 A dilute, liquid solution of low viscosity flows in a laminar fashion past the two-dimensional body sketched below. At the same time, the solute is transferred from the solution to the surface, where the concentration becomes zero. While the kinematic viscosity is low, the diffusion coefficient is considerably lower, that is, the Schmidt number is large.

It has been suggested that an exact asymptotic solution for the mass-transfer problem can be obtained in this case of large Schmidt numbers once the hydrodynamic problem has been solved [1, 2].

(a) Show that the steady-state, mass-transfer, boundary-layer equation applicable in this case can be reduced to an ordinary differential equation by a similarity transformation with the similarity variable

$$\eta = y\sqrt{\beta(x)}\left[9D\int_0^x \sqrt{\beta}\,dx\right]^{-1/3},$$

where β is the velocity derivative at the solid surface

$$\beta(x) = \left.\frac{\partial v_x}{\partial y}\right|_{y=0}.$$

(b) Solve the resulting ordinary differential equation for the concentration as a function of the similarity variable η and verify that Fig. 19.1 gives the concentration profile in the diffusion layer and that the local Nusselt number can be expressed as

$$\mathrm{Nu} = \frac{\ell}{c_\infty}\frac{\partial c_i}{\partial y}\bigg|_{y=0} = \frac{\ell}{\Gamma(4/3)}\sqrt{\beta}\left[9D\int_0^x \sqrt{\beta}\,dx\right]^{-1/3},$$

where ℓ is the radius of curvature at the stagnation point.

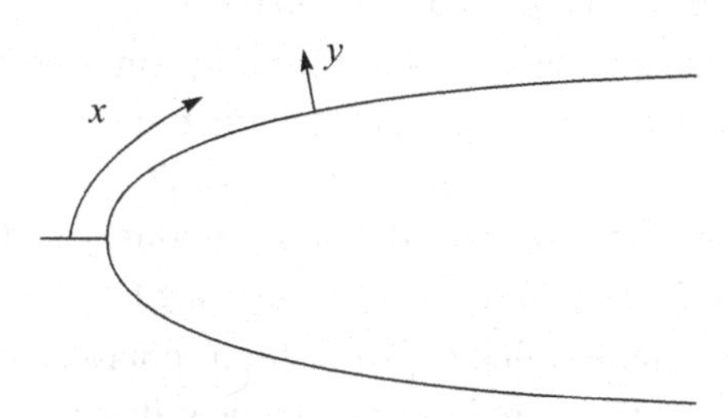

24.3 Consider the problem of mass transfer in two-dimensional laminar boundary layers on a blunt body at low Schmidt numbers. It has been suggested that the concentration profiles can be calculated from the equation

$$U(x)\frac{\partial c_i}{\partial x} - y\frac{dU}{dx}\frac{\partial c_i}{\partial y} = D\frac{\partial^2 c_i}{\partial y^2},$$

where $U(x)$ is the inner limit of the outer Euler solution.

(a) Justify the simplifications which reduce the equation of convective diffusion to this form.

(b) Obtain an exact asymptotic expression for the local mass-transfer rate (for low Sc) by means of a similarity transformation. The following conditions apply:

$$c_i = c_0 \text{ at } y = 0. \quad c_i = c_\infty \text{ at } y = \infty.$$

$U(x)$ is an arbitrary, nonnegative function of x defined in the range $x \geq 0$.

24.4 Under what conditions is the mass-transfer boundary layer much thinner than the hydrodynamic boundary layer?

The Lighthill transformation applies to such a case but is restricted to two-dimensional flows. Carry out a parallel treatment for axisymmetric flows by doing the following:

(a) Show that near the solid surface the velocity profiles can be approximated by

$$v_x = y\beta(x) \text{ and } v_y = -\frac{y^2}{2\mathcal{R}}\frac{d\mathcal{R}\beta}{dx},$$

where $\mathcal{R}(x)$ is the distance of the surface from the axis of symmetry.

(b) Write down the appropriate form of the boundary-layer equation of convective diffusion.

(c) For a constant concentration c_o on the solid surface, determine conclusively whether or not a similarity transformation of the form $(c_i - c_o)/(c_\infty - c_o) = \Theta(\eta)$, where $\eta = y/g(x)$ can be used. If it is possible, you do *not* need to complete the solution by evaluating $\Theta(\eta)$ or $g(x)$.

24.5 An approximate velocity profile for a spherical bubble moving at moderate Reynolds numbers was discussed in Problem 15.2, and the formulation of a mass-transfer problem was considered in Problem 19.3. For small values of the diffusion coefficient, there is a thin diffusion layer near the surface of the bubble.

(a) Discuss qualitatively and quantitatively what determines the order of magnitude of the thickness of the diffusion layer.

(b) Obtain the equation that applies, with the appropriate approximations, in the diffusion layer.

(c) Qualitatively, what approximations are valid in the diffusion layer?

(d) Qualitatively, what approximations are valid in the region outside the diffusion layer?

(e) Predict the manner in which the average Nusselt number depends on the Péclet number $2Rv_\infty/D_i$ for large values of the Péclet number, that is, for small diffusion coefficients.

24.6 Seek a similarity solution to Problem 24.5 for a constant concentration along the surface of the bubble. Be general by formulating diffusion-layer equations for an arbitrary axisymmetric mobile interface where, for small diffusion coefficients, the tangential velocity can be approximated by

$$v_x = v_o(x).$$

Obtain an appropriate approximation to the normal component of the velocity by means of the continuity equation and determine a similarity solution wherein the diffusion-layer thickness is related to $v_0(x)$ and the distance $\mathcal{R}(x)$ of the surface from the axis of symmetry. If you have worked Problem 24.3, compare and contrast this solution for axisymmetric mobile interfaces with that problem for two-dimensional solid surfaces at low Schmidt numbers.

24.7 For the rotating disk and stationary plane in Problem 17.1, the radial velocity distribution near the stationary plane can be expressed as

$$v_r = -Ar\Omega\, y\sqrt{\Omega / \nu}$$

where y is the normal distance from the plane and A is a positive constant.

(a) Use the continuity equation to obtain the normal component v_y of the velocity for distances close to the surface of the plane.

(b) For high Schmidt numbers, determine the distribution of the mass-transfer rate over a mass-transfer section of radius r_0 in the stationary plane. The surface of the plane outside r_0 is insulating. The concentration at the active surface is c_0, and that of the bulk of the fluid is c_∞. You are expected to use a diffusion-layer approach.

(c) What are the principal differences between the mass-transfer distribution on the stationary plane and that which would result on an active section of radius r_0 on the surface of the disk rotating above the plane?

24.8 Show how the Mangier transformation (see Problem 23.2) can be used to transform the problem for an axisymmetric diffusion layer into one for a two-dimensional diffusion layer.

24.9 Consider mass transfer to an axisymmetric body at high Schmidt numbers (see also Problem 24.4).

(a) Show that near the surface the velocity profiles can be approximated by

$$v_x = y\beta(x) \quad \text{and} \quad v_y = -\frac{y^2}{2\mathcal{R}}\frac{d\mathcal{R}\beta}{dx}.$$

(b) Use the similarity variable

$$\eta = y\sqrt{\mathcal{R}\beta} \Big/ \left[9D\int_0^x \mathcal{R}\sqrt{\mathcal{R}\beta}\, dx \right]^{1/3}$$

to solve for the concentration profile in the diffusion layer at high Schmidt numbers for a constant surface concentration, and use this result to obtain the local and average Nusselt numbers.

(c) Show how the treatment of mass transfer to a sphere in Stokes flow at high Péclet numbers can be considered a special case of this transformation.

(d) Show how the treatment of the rotating disk at high Schmidt numbers can be considered a special case of this transformation.

(e) Use the Mangier transformation (see Problem 24.8) and the results of Problem 24.2 to obtain the local Nusselt number. This should agree with the result in Part (b).

24.10 Consider mass transfer from the solid wall to a fluid in fully developed laminar flow in a circular tube. Downstream of z = 0, the flux at the wall is constant. The Péclet number is high, and we are interested in the concentration variation near the beginning of the mass-transfer section.

(a) Write down the equation of convective diffusion in the form that you intend to solve. For each term omitted in simplifying the equation

$$\frac{\partial c_i}{\partial t} + \mathbf{v} \cdot \nabla c_i = D\nabla^2 c_i$$

mention briefly the justification.

(b) A similarity transformation will work for the simplified problem. Demonstrate this by finding a similarity variable and an ordinary differential equation for a function of this variable and by showing how the boundary conditions are to be satisfied.

(c) Solve for the concentration distribution in the region in question. Obtain an expression for the local Nusselt number as a function of Péclet number and position on the wall.

24.11 The Lighthill transformation provides us with a useful solution to the diffusion-layer equation of convective diffusion. Let us investigate whether this can be extended easily.

(a) Write down the boundary-layer form of the equation of convective diffusion. What should subsequently be assumed in order to obtain the solution by means of the Lighthill transformation?

(b) For a laminar boundary-layer problem, show how to obtain the second derivative of the velocity at the wall

$$\gamma = \left.\frac{\partial^2 v_x}{\partial y^2}\right|_{y=0}$$

in terms of the pressure distribution in the boundary layer.

(c) Express the velocity profiles in Taylor series in distance from the wall. Include the terms involving γ, and do this for a general axisymmetric geometry.

(d) Transform the diffusion-layer problem from one involving the coordinates x and y to one involving the coordinates x and ξ, where ξ is the Lighthill similarity variable

$$\xi = \frac{y}{g(x)} = \frac{y\sqrt{\mathcal{R}\beta}}{\left[9D\int_0^x \mathcal{R}\sqrt{\mathcal{R}\beta}\, dx\right]^{1/3}}$$

(e) This coordinate system should put us in a good position to assess how easy it would be to extend the Lighthill solution. Express the concentration as

$$\Theta = \frac{c_i - c_0}{c_\infty - c_0} = \Theta_0(\xi) + \Theta_1(x, \xi) + \cdots$$

and formulate the problem for Θ_1 as the second term in an expansion for small diffusion coefficient D. You need not attempt to determine whether there is any simple solution for Θ_1, but this is the direction that the investigation would take next.

References

1. M. J. Lighthill. "Contributions to the theory of heat transfer through a laminar boundary layer." *Proceedings of the Royal Society*, **A202**, 359–377 (1950).
2. Andreas Acrivos. "Solution of the Laminar Boundary Layer Energy Equation at High Prandtl Numbers." *The Physics of Fluids*, **3**, 657–658 (1960).

Chapter 25

Blasius Series for Mass Transfer

In Chapter 23, we saw how the boundary-layer velocity profiles could be expanded in a power series near the stagnation point of a symmetric cylinder. We wrote

$$U(x) = u_1 x + u_3 x^3 + u_5 x^5 + \cdots, \tag{25.1}$$

$$v_x(x,y) = u_1 x f_1'(\eta) + 4u_3 x^3 f_3'(\eta) + \cdots, \tag{25.2}$$

where $\eta = y\sqrt{u_1 / \nu}$. In a similar manner, the concentration can be expanded in a power series (see, for example, Refs. [1, 2]):

$$\Theta(x,y) = \frac{c_i - c_o}{c_\infty - c_o} = \Theta_o(\eta) + \frac{u_3}{u_1} x^2 \Theta_2(\eta) + \frac{u_5}{u_1} x^4 \Theta_4(\eta) + \cdots. \tag{25.3}$$

The functions Θ_o and Θ_2 are "universal functions" independent of u_1, u_3, etc., although they depend on the Schmidt number. In order to form universal functions, the higher-order coefficients must be broken up in a manner similar to the hydrodynamic functions (compare Eq. 23.6):

The Newman Lectures on Transport Phenomena
John Newman and Vincent Battaglia

ISBN 978-981-4774-27-7 (Hardcover), 978-1-315-10829-2 (eBook)
www.jennystanford.com

$$\left.\begin{aligned}
\Theta_4 &= a_4 + \frac{u_3^2}{u_1 u_5} b_4, \\
\Theta_6 &= a_6 + \frac{u_3 u_5}{u_1 u_7} b_6 + \frac{u_3^3}{u_1^2 u_7} d_6, \\
\Theta_8 &= a_8 + \frac{u_3 u_7}{u_1 u_9} b_8 + \frac{u_5^2}{u_1 u_9} d_8 + \frac{u_3^2 u_5}{u_1^2 u_9} e_8 + \frac{u_3^4}{u_1^3 u_9} p_8, \\
\Theta_{10} &= a_{10} + \frac{u_3 u_9}{u_1 u_{11}} b_{10} + \frac{u_5 u_7}{u_1 u_{11}} d_{10} + \frac{u_3^2 u_7}{u_1^2 u_{11}} e_{10} \\
&\quad + \frac{u_3 u_5^2}{u_1^2 u_{11}} p_{10} + \frac{u_3^3 u_5}{u_1^3 u_{11}} r_{10} + \frac{u_3^5}{u_1^4 u_{11}} s_{10}.
\end{aligned}\right\} \quad (25.4)$$

The differential equations for some of the universal mass-transfer, boundary-layer functions are

$$\left.\begin{aligned}
\frac{1}{\text{Sc}}\Theta_o'' + f_1\Theta_o' &= 0, \\
\frac{1}{\text{Sc}}\Theta_2'' + f_1\Theta_2' - 2f_1'\Theta_2 &= -12 f_3\Theta_o', \\
\frac{1}{\text{Sc}}a_4'' + f_1 a_4' - 4f_1' a_4 &= -30 g_5\Theta_o', \\
\frac{1}{\text{Sc}}b_4'' + f_1' b_4 - 4f_1' b_4 &= -12 f_3\Theta_2' + 8f_3'\Theta_2 - 30 h_5\Theta_o',
\end{aligned}\right\} \quad (25.5)$$

and the corresponding boundary conditions are

$$\left.\begin{aligned}
\Theta_o = \Theta_2 = a_4 = b_4 = 0 \ \text{at} \ \eta = 0. \\
\Theta_o = 1, \Theta_2 = a_4 = b_4 = 0 \ \text{at} \ \eta = \infty.
\end{aligned}\right\} \quad (25.6)$$

The functions f_1, f_3, g_5, and h_5 appearing in Eqs. 25.5 are the same universal functions considered in Chapter 23.

Since the universal mass-transfer, boundary-layer functions depend on Sc as well as on η, their tabulation could become unwieldy. The local mass-transfer rate is of interest and is given by

$$\begin{aligned}
N_y(x) = -D\left.\frac{\partial c_i}{\partial y}\right|_{y=0} &= -D(c_\infty - c_o)\sqrt{\frac{u_1}{\nu}}\left\{\Theta_0'(0) + \frac{u_3}{u_1}x^2\Theta_2'(0)\right. \\
&\quad \left. + \frac{u_5}{u_1}x^4\Theta_4'(0) + \frac{u_7}{u_1}x^6\Theta_6'(0) + \cdots\right\}.
\end{aligned} \quad (25.7)$$

The coefficients necessary for the calculation of the rate of mass transfer are given in Table 25.1. Asymptotic forms for large and small Schmidt numbers are given in Tables 25.2 and 25.3, respectively.

Table 25.1 Dimensionless mass-transfer coefficients from Blasius series.

Sc	$\Theta_0'(0)$	$\Theta_2'(0)$	$a_4'(0)$	$b_4'(0)$	$a_6'(0)$	$b_6'(0)$	$d_6'(0)$
0.005	0.0545	0.0424	0.0492	-0.0114	0.0539	-0.0264	0.0083
0.01	0.0760	0.0598	0.0704	-0.0171	0.0781	-0.0414	0.0138
0.02	0.1054	0.0844	0.1009	-0.0260	0.1134	-0.0652	0.0228
0.05	0.1610	0.1323	0.1619	-0.0450	0.1856	-0.1177	0.0433
0.10	0.2195	0.1847	0.2304	-0.0675	0.2678	-0.1808	0.0682
0.20	0.2964	0.2557	0.3250	-0.0996	0.3828	-0.2718	0.1042
0.50	0.4334	0.3867	0.5025	-0.1613	0.6007	-0.4480	0.1742
0.70	0.4959	0.4476	0.5860	-0.1906	0.7036	-0.5320	0.2076
1.00	0.5705	0.5210	0.6868	-0.2263	0.8284	-0.6343	0.2482
2.00	0.7437	0.6932	0.9247	-0.3107	1.1235	-0.8773	0.3451
5.00	1.0434	0.9938	1.3419	-0.4596	1.6428	-1.3077	0.5170
10.00	1.3388	1.2912	1.7557	-0.6078	2.1588	-1.7374	0.6891
20.00	1.7104	1.6657	2.2773	-0.7951	2.8099	-2.2812	0.9073
50.00	2.3529	2.3132	3.1798	-1.1198	3.9368	-3.2249	1.2865
100.00	2.9869	2.9519	4.0700	-1.4404	5.0486	-4.1578	1.6616
200.00	3.7854	3.7557	5.1901	-1.8444	6.4476	-5.3337	2.1345
500.00	5.1685	5.1456	7.1267	-2.5439	8.8664	-7.3717	2.9536
1000.00	6.5352	6.5163	9.0358	-3.2352	11.2504	-9.3876	3.7629

Sc	$a_8'(0)$	$b_8'(0)$	$d_8'(0)$	$e_8'(0)$	$p_8'(0)$
0.005	0.0578	-0.0304	-0.0160	0.0339	-0.0110
0.01	0.0846	-0.0488	-0.0263	0.0582	-0.0193
0.02	0.1242	-0.0784	-0.0429	0.0981	-0.0330
0.05	0.2064	-0.1444	-0.0804	0.1889	-0.0640
0.10	0.3011	-0.2241	-0.1259	0.2994	-0.1017
0.20	0.4345	-0.3393	-0.1920	0.4593	-0.1560
0.50	0.6889	-0.5629	-0.3202	0.7693	-0.2611

Sc	$a_8'(0)$	$b_8'(0)$	$d_8'(0)$	$e_8'(0)$	$p_8'(0)$
0.70	0.8096	−0.6697	−0.3815	0.9176	−0.3114
1.00	0.9561	−0.7999	−0.4563	1.0984	−0.3727
2.00	1.3033	−1.1098	−0.6344	1.5295	−0.5192
5.00	1.9156	−1.6597	−0.9506	2.2971	−0.7805
10.00	2.5250	−2.2097	−1.2672	3.0671	−1.0432
20.00	3.2944	−2.9065	−1.6684	4.0446	−1.3771
50.00	4.6268	−4.1167	−2.3655	5.7451	−1.9586
100.00	5.9414	−5.3137	−3.0551	7.4283	−2.5346
200.00	7.5959	−6.8231	−3.9249	9.5514	−3.2614
500.00	10.4563	−9.4404	−5.4333	13.2311	−4.5211
1000.00	13.2758	−12.0308	−6.9263	16.8690	−5.7662

Sc	$a_{10}'(0)$	$b_{10}'(0)$	$d_{10}'(0)$	$e_{10}'(0)$	$p_{10}'(0)$	$r_{10}'(0)$	$s_{10}'(0)$
0.005	0.0528	−0.0344	−0.0383	0.0429	0.0488	−0.0657	0.0189
0.01	0.0849	−0.0561	−0.0642	0.0743	0.0850	−0.1157	0.0334
0.02	0.1315	−0.0912	−0.1066	0.1260	0.1448	−0.1982	0.0571
0.05	0.2251	−0.1696	−0.2024	0.2432	0.2806	−0.3841	0.1104
0.10	0.3317	−0.2647	−0.3191	0.3856	0.4457	−0.6089	0.1744
0.20	0.4822	−0.4022	−0.4884	0.5914	0.6840	−0.9325	0.2664
0.50	0.7705	−0.6693	−0.8175	0.9905	1.1463	−1.5589	0.4443
0.70	0.9076	−0.7970	−0.9751	1.1815	1.3676	−1.8587	0.5294
1.00	1.0743	−0.9528	−1.1674	1.4145	1.6377	−2.2249	0.6335
2.00	1.4699	−1.3240	−1.6260	1.9709	2.2829	−3.1003	0.8825
5.00	2.1689	−1.9838	−2.4420	2.9630	3.4340	−4.6650	1.3285
10.00	2.8653	−2.6446	−3.2599	3.9594	4.5908	−6.2401	1.7780
20.00	3.7451	−3.4822	−4.2975	5.2252	6.0609	−8.2442	2.3506
50.00	5.2692	−4.9380	−6.1017	7.4286	8.6207	−11.7371	3.3494
100.00	6.7733	−6.3784	−7.8875	9.6107	11.1562	−15.1990	4.3399
200.00	8.6665	−8.1955	−10.1406	12.3638	14.3556	−19.5686	5.5905
500.00	11.9396	−11.3471	−14.0492	17.1370	19.9031	−27.1456	7.7593
1000.00	15.1658	−14.4676	−17.9196	21.8578	25.3901	−34.6385	9.9040

Table 25.2 Asymptotes for large Schmidt numbers.

$\Theta_0'(0) = 0.6608\ \text{Sc}^{1/3}$
$\Theta_2'(0) = 0.6658\ \text{Sc}^{1/3}$
$a_4'(0) = 0.9280\ \text{Sc}^{1/3}$
$b_4'(0) = 0.3339\ \text{Sc}^{1/3}$
$a_6'(0) = 1.1592\ \text{Sc}^{1/3}$
$b_6'(0) = -0.9719\ \text{Sc}^{1/3}$
$d_6'(0) = 0.3917\ \text{Sc}^{1/3}$
$a_8'(0) = 1.3711\ \text{Sc}^{1/3}$
$b_8'(0) = -1.2475\ \text{Sc}^{1/3}$
$d_8'(0) = -0.7190\ \text{Sc}^{1/3}$
$e_8'(0) = 1.7590\ \text{Sc}^{1/3}$
$p_8'(0) = -0.6028\ \text{Sc}^{1/3}$
$a_{10}'(0) = 1.5690\ \text{Sc}^{1/3}$
$b_{10}'(0) = -1.5016\ \text{Sc}^{1/3}$
$d_{10}'(0) = -1.8628\ \text{Sc}^{1/3}$
$e_{10}'(0) = 2.2556\ \text{Sc}^{1/3}$
$p_{10}'(0) = 2.6516\ \text{Sc}^{1/3}$
$r_{10}'(0) = -3.6335\ \text{Sc}^{1/3}$
$s_{10}'(0) = 1.0506\ \text{Sc}^{1/3}$

Table 25.3 Asymptotes for small Schmidt numbers.

$\Theta_0'(0) = 0.7979\ \text{Sc}^{1/2}$
$\Theta_2'(0) = 0.5984\ \text{Sc}^{1/2}$
$a_4'(0) = 0.6649\ \text{Sc}^{1/2}$
$b_4'(0) = -0.1247\ \text{Sc}^{1/2}$
$a_6'(0) = 0.6981\ \text{Sc}^{1/2}$
$b_6'(0) = -0.2327\ \text{Sc}^{1/2}$
$d_6'(0) = 0.0436\ \text{Sc}^{1/2}$
$a_8'(0) = 0.7181\ \text{Sc}^{1/2}$
$b_8'(0) = -0.2244\ \text{Sc}^{1/2}$
$d_8'(0) = -0.0997\ \text{Sc}^{1/2}$
$e_8'(0) = 0.1122\ \text{Sc}^{1/2}$
$p_8'(0) = -0.0175\ \text{Sc}^{1/2}$
$a_{10}'(0) = 0.7314\ \text{Sc}^{1/2}$
$b_{10}'(0) = -0.2194\ \text{Sc}^{1/2}$
$d_{10}'(0) = -0.1828\ \text{Sc}^{1/2}$
$e_{10}'(0) = 0.1029\ \text{Sc}^{1/2}$
$p_{10}'(0) = 0.0914\ \text{Sc}^{1/2}$
$r_{10}'(0) = -0.0571\ \text{Sc}^{1/2}$
$s_{10}'(0) = 0.0075\ \text{Sc}^{1/2}$

Problem

25.1 Apply the result of Problem 24.2 to obtain the high-Sc asymptotic form of the mass-transfer rate as given by the Blasius series for mass transfer, that is, verify Table 25.2. Calculate only $\Theta_0'(0)$, $\Theta_2'(0)$, $a_4'(0)$, and $b_4'(0)$.

References

1. Hermann Schlichting. *Boundary-Layer Theory*, translated by J. Kestin. New York: McGraw-Hill Book Company, 1979, p. 303.
2. John Newman. "Blasius Series for Heat and Mass Transfer." *International Journal of Heat and Mass Transfer*, **9**, 705–709 (1966).

Chapter 26

Graetz–Nusselt–Lévêque Problem

An important problem that received early analytic treatment is that of mass transfer to the wall of a tube in which Poiseuille flow is presumed to prevail:

$$v_z = 2 < v_z > \left(1 - \frac{r^2}{R^2}\right), \tag{26.1}$$

$$\mathbf{v_r} = v_\theta = 0. \tag{26.2}$$

Although the Reynolds number can attain values of 2000 before the flow becomes turbulent, this is not a boundary-layer flow.

The equation of convective diffusion is

$$v_z \frac{\partial c_i}{\partial z} = D\left[\frac{1}{r}\frac{\partial}{\partial r}\left(r\frac{\partial c_i}{\partial r}\right) + \frac{\partial^2 c_i}{\partial z^2}\right]. \tag{26.3}$$

This is usually solved for mass transfer to a section with a constant wall concentration,

$$c_i = c_0 \quad \text{at} \quad r = R, \tag{26.4}$$

beginning at $z = 0$ after the Poiseuille flow is fully developed. For other boundary conditions, we may state

$$c_i = c_0 \text{ at } z = 0 \quad \text{and} \quad \partial c_i/\partial r = 0 \text{ at } r = 0. \tag{26.5}$$

The Newman Lectures on Transport Phenomena
John Newman and Vincent Battaglia

ISBN 978-981-4774-27-7 (Hardcover), 978-1-315-10829-2 (eBook)
www.jennystanford.com

Let us introduce dimensionless variables

$$\xi = \frac{r}{R}, \quad \Theta = \frac{c_i - c_o}{c_b - c_o}, \quad \zeta = \frac{zD}{2 < v_z > R^2}. \tag{26.6}$$

The equation of convective diffusion becomes

$$(1-\xi^2)\frac{\partial \Theta}{\partial \zeta} = \frac{1}{\xi}\frac{\partial}{\partial \xi}\left(\xi \frac{\partial \Theta}{\partial \xi}\right) + \frac{1}{\mathrm{Pe}^2}\frac{\partial^2 \Theta}{\partial \zeta^2}, \tag{26.7}$$

where

$$\mathrm{Pe} = \mathrm{Re} \cdot \mathrm{Sc} = \frac{2R < v_z >}{D} = \frac{2R < v_z >}{\nu}\frac{\nu}{D} \tag{26.8}$$

is the Péclet number. On the assumption that the Péclet number is large, we discard the second derivative with respect to ζ:

$$(1-\xi^2)\frac{\partial \Theta}{\partial \zeta} = \frac{\partial^2 \Theta}{\partial \xi^2} + \frac{1}{\xi}\frac{\partial \Theta}{\partial \xi}, \tag{26.9}$$

$$\Theta = 1 \text{ at } \zeta = 0.$$
$$\Theta = 0 \text{ at } \xi = 1.$$
$$\partial \Theta / \partial \xi = 0 \text{ at } \xi = 0.$$

The total amount of material transferred to the wall in a length L is

$$J = -\int_0^L D \left.\frac{\partial c_i}{\partial r}\right|_{r=R} 2\pi R dz \tag{26.10}$$

or

$$\frac{J}{\pi R^2 (c_b - c_o) < v_z >} = 2\frac{\mathrm{Nu}}{\mathrm{Pe}}\frac{L}{R} = -4\int_0^z \left.\frac{\partial \Theta}{\partial \xi}\right|_{\xi=1} d\zeta, \tag{26.11}$$

where

$$z = \frac{DL}{2 < v_z > R^2}. \tag{26.12}$$

26.1 Solution by Separation of Variables

Graetz (1885), followed by Nusselt (1910), treated this problem by the method of separation of variables [1, 2]:

$$\Theta = \sum_{k=1}^{\infty} A_k e^{-\lambda_k^2 \zeta} R_k(\xi), \tag{26.13}$$

in which R_k satisfies the equation

$$\frac{1}{\xi}\frac{d}{d\xi}\left(\xi \frac{dR_k}{d\xi}\right) + \lambda_k^2 (1-\xi^2) R_k = 0, \tag{26.14}$$

with the boundary conditions

$$\left.\begin{array}{r} R_k = 0 \ \text{ at } \ \xi = 1. \\ dR_k/d\xi = 0 \ \text{ at } \ \xi = 0. \end{array}\right\} \tag{26.15}$$

The solution to this Sturm–Liouville system has been calculated, and R_1, R_2, and R_3 are tabulated by Jakob (*Heat Transfer*, vol. 1, p. 455) [3].

The total amount of material transferred to the wall can be calculated from the expression

$$1 - \frac{J}{\pi R^2 (c_b - c_o) < v_z >} = \sum_{k=1}^{\infty} 4A_k e^{-\lambda_k^2 z} \int_0^1 \xi (1-\xi^2) R_k(\xi) d\xi$$
$$= \sum_{k=1}^{\infty} M_k e^{-\lambda_k^2 z}, \tag{26.16}$$

where the values of M_k and λ_k are given for ten terms in Table 26.1. The eigenvalues λ_k for the first five terms are taken from Abramowitz (1953) [4], and the values of M_k for these terms were calculated from the results of Lipkis (1956) [5]. Values for the higher terms were calculated from the formulas of Sellars, Tribus, and Klein (1956) [6], which formulas are valid in the asymptotic limit of large λ_k:

$$\lambda_k = 4k - \frac{4}{3}, \tag{26.17}$$

$$M_k = 8C\lambda_k^{-7/3}, \tag{26.18}$$

where

$$C = \frac{3^{1/6} 18^{1/3}}{\pi} \frac{\Gamma(5/3)}{\Gamma(4/2)} = 1.012787288. \tag{26.19}$$

Table 26.1 Values of λ_k and M_k for the Graetz series.*

k	λ_k	M_k
1	2.7043644	0.8190504
2	6.6790315	0.0975269
3	10.6733795	0.0325040
4	14.6710785	0.0154402
5	18.6698719	0.0087885
6	22.6691434	0.0055838
7	26.6686620	0.0038202
8	30.6683233	0.0027564
9	34.6680738	0.0020702
10	38.6678834	0.0016043

*Revised (December 1968) to reflect the developments reported in Ref. [7].

26.2 Solution for Very Short Distances

For very short distances, Lévêque (1928) [8] recognized that there is a diffusion layer near the wall, and derivatives with respect to ξ become large. Within the diffusion layer, the following approximations apply:

$$1-\xi^2 = 2y - y^2 \approx 2y \quad \text{and} \quad \frac{1}{\xi}\frac{\partial \Theta}{\partial \xi} << \frac{\partial^2 \Theta}{\partial \xi^2},$$

where $y = 1 - \xi$, and the diffusion equation becomes

$$2y\frac{\partial \Theta}{\partial \zeta} = \frac{\partial^2 \Theta}{\partial y^2} \tag{26.20}$$

with boundary conditions

$$\Theta = 0 \text{ at } y = 0 \quad \text{and} \quad \Theta \to 1 \text{ as } y \to \infty. \tag{26.21}$$

The similarity transformation

$$\eta = y\,(2/9\zeta)^{1/3} \tag{26.22}$$

reduces the diffusion equation to an ordinary differential equation

$$\Theta'' + 3\eta^2\Theta' = 0, \tag{26.23}$$

with the solution

$$\Theta = \frac{1}{\Gamma(4/3)} \int_0^{\eta} e^{-x3} dx . \qquad (26.24)$$

The total amount of material transferred to the wall is given by

$$\frac{J}{\pi R^2 (c_b - c_o) < v_z >} = \frac{3\sqrt{48}}{\Gamma(4/3)} Z^{2/3} = 4.070\, Z^{2/3}. \qquad (26.25)$$

26.3 Extension of Lévêque Solution

It should be possible to calculate corrections to the concentration profile based on the Lévêque solution (see Problem 26.1 and Ref. [9]). On this basis, the average Nusselt number referred to the concentration difference at the inlet can be expressed as

$$\mathrm{Nu} = 1.6151 \left(\frac{\mathrm{ScRe}}{L/2R} \right)^{1/3} - 1.2 - 0.28057 \left(\frac{L/2R}{\mathrm{ScRe}} \right)^{1/3} + O\left(\frac{L/2R}{\mathrm{ScRe}} \right)^{2/3} . \qquad (26.26)$$

From the Graetz series and the Lévêque series, the values of J can be calculated to better than 0.1% for all values of Z. The result is shown in Fig. 26.1.

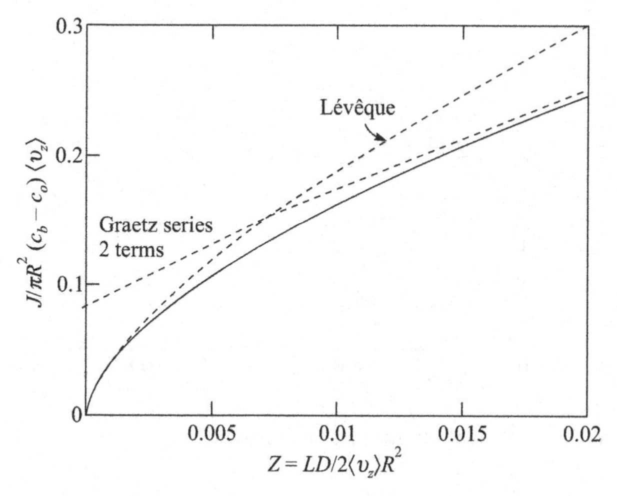

Figure 26.1 Rate of mass transfer to the wall in laminar tube flow for a length L.

26.4 Mass Transfer in Annuli

The Lévêque solution is useful, as indicated by the above solution, for mass-transfer lengths such that $L/2R$ << ScRe. Frequently, this covers the entire range of interest, particularly for liquids where the Schmidt number is large. It is straightforward to apply the method of Lévêque to mass transfer in annular spaces by using the velocity derivative at the walls of the annulus instead of that for the tube, and this has been done by Friend and Metzner [10]. The result for the average Nusselt number is

$$\text{Nu} = 1.6151\left(\frac{\text{ReSc}}{L/d_e}\right)^{1/3}\phi^{1/3}, \qquad (26.27)$$

where $\text{Re} = d_e \langle v_z \rangle / \nu$, $d_e = 2R(1-\kappa)$ is the equivalent diameter, and κ is the ratio of the inner radius to the outer radius. The function $\phi^{1/3}$ is plotted against κ in Fig. 26.2 for both the inner and the outer walls.

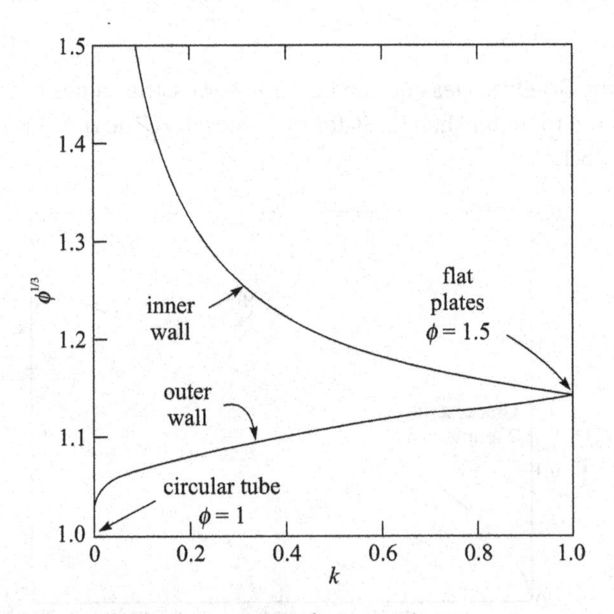

Figure 26.2 Coefficient for mass transfer in annuli.

Problems

26.1 For mass transfer in Poiseuille flow in a pipe, we found that if axial diffusion is neglected, the mass-transfer rate could be expressed as

$$\frac{J}{\pi R^2(c_b - c_o) < v_z >} = \frac{(48)^{1/3}}{\Gamma(4/3)} Z^{2/3} = 4.070\, Z^{2/3},$$

where

$$Z = \frac{DL}{2 < v_z > R^2},$$

J is the total amount of material transferred to the wall in a length L, c_b is the inlet concentration, and c_o is the (constant) wall concentration. This is the first approximation for small values of Z as obtained by Lévêque. Show how to use the Lévêque solution as a basis for obtaining higher-order corrections to the mass-transfer rate in an expansion valid for small values of Z, while continuing to neglect axial diffusion. If you reduce the problem to a set of ordinary differential equations, you are not expected to solve them, but you should indicate the form of the final expression for the mass-transfer rate even though you do not evaluate the constants. Indicate any possible weak points in the development.

26.2 For mass transfer in laminar pipe flow, we reduced the equation of convective diffusion to the form

$$(1-\xi^2)\frac{\partial \Theta}{\partial \zeta} = \frac{1}{\xi}\frac{\partial}{\partial \xi}\left(\xi \frac{\partial \Theta}{\partial \xi}\right) + \frac{1}{\mathrm{Pe}^2}\frac{\partial^2 \Theta}{\partial \zeta^2}$$

where $\mathrm{Pe} = 2R\langle v_z \rangle / D$, $\xi = r/R$, $\Theta = (c_i - c_o)/(c_b - c_o)$, and $\zeta = zD/2\langle v_z \rangle R^2$. With the neglect of axial diffusion, we investigated mass transfer with the Graetz series and, close to the beginning of the mass-transfer section, by the method of Lévêque.

Now we want to investigate the effect of the axial diffusion terms, still with a large Péclet number. For this purpose, we let the boundary conditions take the more reasonable form

$$\partial\Theta/\partial\xi = 0 \text{ at } \xi = 1,\ \zeta < 0, \text{an insulating wall.}$$
$$\Theta = 0 \text{ at } \xi = 1,\ \zeta > 0, \text{a constant wall concetration.}$$
$$\partial\Theta/\partial\xi = 0 \text{ at } \xi = 0.$$
$$\Theta \to 1 \text{ as } \zeta \to -\infty.$$
$$\Theta \to 0 \text{ as } \zeta \to +\infty.$$

On the basis of our previous work with the problem, we expect that the axial diffusion terms are important only near the beginning of the mass-transfer section. Use the Lévêque solution to find the region where axial diffusion should become important. On this basis, find stretched coordinates for treating this region so that one can write the applicable differential equation in a form appropriate to this region but without the explicit appearance of the Péclet number. Formulate boundary conditions for the problem in this region. These should also be independent of the Péclet number, so that one has a well-posed, singular-perturbation problem for this region where axial diffusion is important.

26.3 A flow-through porous electrode can be approximated by a metal block with a large number of straight, parallel, circular holes of small radius (see figure). A solution flows through these pores with a superficial velocity v_0 (ratio of volumetric flow rate to total electrode cross-sectional area), and a reaction can occur at the walls of the pores. In the analysis of this system, one wants to use a one-dimensional approximation and assume that the concentration is uniform across the cross section of a pore. This simplifies the analysis because one then needs to determine the current, potential, concentration, and reaction distributions only as functions of distance through the electrode, and the radial dependence in the pore is ignored.

To test the validity of this assumption, use the results of the Graetz–Nusselt–Lévêque problem to obtain an expression for the axial distance required for the average concentration in the pore to reach 90% of the wall concentration after a step in wall concentration from zero to c_1. You need not distinguish between the average concentration and the "cup-mixing" concentration.

If the distance so calculated is small compared to the distance over which most of the reaction occurs, the assumption would appear to be justified.

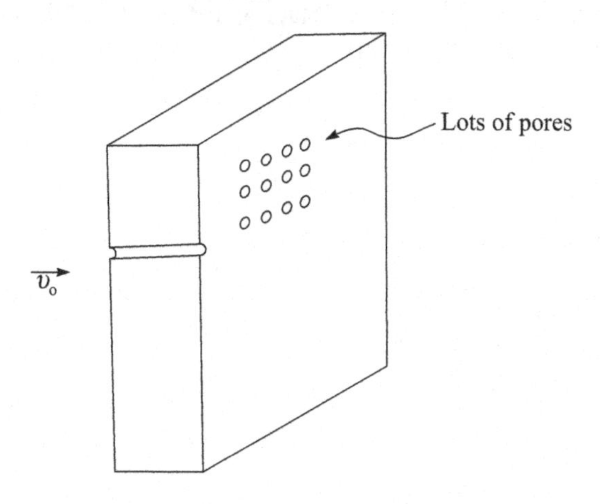

References

1. L. Graetz. "Ueber die Wärmeleitungsfähigkeit von Flüssig-keiten." *Annalen der Physik nd Chemie*, **18**, 79–94 (1883), **25**, 337–357 (1885).
2. Wilhelm Nusselt. "Die Abhängigkeit der Wärmeübergangszahl von der Rohrlänge." *Zeitschrift des Vereines-deutscher Ingenieure*, **54**, 1154–1158 (1910).
3. Max Jakob. *Heat Transfer*. Volume I. New York: John Wiley & Sons, Inc., 1949. pp. 451–464.
4. Milton Abramowitz. "On the Solution of the Differential Equation Occurring in the Problem of Heat Convection in Laminar Flow Through a Tube." *Journal of Mathematics and Physics*, **32**, 184–187 (1953).
5. R. P. Lipkis. "Discussion." *Transactions of the American Society of Mechanical Engineers*, **78**, 447–446 (1956).
6. J. R. Sellars, Myron Tribus, and J. S. Klein. "Heat Transfer to Laminar Flow in a Round Tube or Flat Conduit: The Graetz Problem Extended." *Transactions of the American Society of Mechanical Engineers*, **78**, 441–447 (1956).
7. John Newman. "The Graetz Problem." UCRL-18646. January, 1969.
8. M. A. Lévêque. "Les Lois de la Transmission de Chaleur par Convection." *Annales des Mines, Memoires*, ser. 12, **13**, 201–299, 305–362, 381–415 (1928).

9. John Newman. "Extension of the Lévêque Solution." *Journal of Heat Transfer*, **91**, 177–178 (1969).
10. W. L. Friend and A. B. Metzner. "Turbulent Heat Transfer Inside Tubes and the Analogy Among Heat, Mass, and Momentum Transfer." *AIChE. Journal*, **4**, 393–402 (1958).

Chapter 27

Natural Convection

Natural convection refers to the situation when the convective fluid motion is due to density differences generated as part of processes of heat transfer or mass transfer. It is important when other sources of fluid motion are absent or weak. Since convection plays an important role in the processes of convective diffusion, the situation is now essentially more complex than in the case of forced convection. The mass-transfer process produces the fluid motion; therefore, the rate of mass transfer is no longer linear in the concentration difference between the bulk fluid and the solid surface since the degree of motion also depends on this difference through the density difference.

The momentum equation 13.3 is now coupled to the mass-transfer process. In the simplest treatment, we should like to assume that the properties are constant. However, since the fluid motion is due to variations in density, it is necessary to account for the important aspects of this variation. Expand the density about its value at $\bar{c}_i$:

$$\rho = \bar{\rho} + \bar{\rho}\,\bar{\beta}\,(c_i - \bar{c}_i) + \cdots, \tag{27.1}$$

where

$$\bar{\beta} = \frac{1}{\rho}\left(\frac{\partial \rho}{\partial c_i}\right)_{T,p}\Bigg|_{c_i=\bar{c}_i}. \tag{27.2}$$

The Newman Lectures on Transport Phenomena
John Newman and Vincent Battaglia

ISBN 978-981-4774-27-7 (Hardcover), 978-1-315-10829-2 (eBook)
www.jennystanford.com

The nature of this expansion suggests a restriction to binary systems at uniform temperature and a negligible dependence of density on pressure. The momentum equation becomes

$$\rho\left(\frac{\partial \mathbf{v}}{\partial t}+\mathbf{v}\cdot\nabla\mathbf{v}\right)=-\nabla p+\mu\nabla^2\mathbf{v}+\bar{\rho}\mathbf{g}+\bar{\rho}\bar{\beta}\mathbf{g}(c_i-\bar{c}_i). \quad (27.3)$$

It is occasionally justified to assume that the pressure variation is given by

$$\nabla p=\bar{\rho}\mathbf{g}, \quad (27.4)$$

in which case the momentum equation becomes

$$\bar{\rho}\left(\frac{\partial \mathbf{v}}{\partial t}+\mathbf{v}\cdot\nabla\mathbf{v}\right)=\mu\nabla^2\mathbf{v}+\bar{\rho}\bar{\beta}\mathbf{g}(c_i-\bar{c}_i). \quad (27.5)$$

This applies only when the buoyancy effect is the only source of motion. This assumption can be justified, for example, for a boundary layer where the pressure variation across the diffusion layer is small and the external fluid is stagnant and has the bulk density $\bar{\rho}$.

In general, it is not necessary to assume Eq. 27.4, and, in fact, without this assumption, there are already the same number of equations as unknowns. The use of the assumption should be justified in each particular case.

The equations of continuity 13.2 and of connective diffusion 13.6 still apply since variations in density are assumed to be small.

The important dimensionless groups of forced convection, Re and Sc, are replaced in free convection by the Grashof number and the Schmidt number

$$\text{Gr}=\frac{\bar{\rho}^2\bar{\beta}g\Delta c_i\ell^3}{\mu^2}=\frac{\bar{\rho}g\Delta\rho\ell^3}{\mu^2}, \quad \text{Sc}=\frac{\nu}{D}. \quad (27.6)$$

A characteristic velocity for free convection comes not, of course, from any external flow, but from the driving force for fluid motion. For example,

$$U=\sqrt{Lg\bar{\beta}\Delta c_i}=\sqrt{L\Delta\rho g/\bar{\rho}}. \quad (27.7)$$

For large values of the Grashof number, the concepts of a boundary layer still apply. For a two-dimensional body and $\overline{c}_i = c_\infty$, the steady boundary-layer equations become

$$\upsilon_x \frac{\partial \upsilon_x}{\partial x} + \upsilon_y \frac{\partial \upsilon_x}{\partial y} = \nu \frac{\partial^2 \upsilon_x}{\partial y^2} + \overline{\beta} g_x (c_i - c_\infty), \tag{27.8}$$

$$\upsilon_x \frac{\partial c_i}{\partial x} + \upsilon_y \frac{\partial c_i}{\partial y} = D \frac{\partial^2 c_i}{\partial y^2}, \tag{27.9}$$

$$\frac{\partial \upsilon_x}{\partial x} + \frac{\partial \upsilon_y}{\partial y} = 0, \tag{27.10}$$

where the boundary-layer coordinates are shown in Fig. 27.1. Since $g_x = -g \sin \epsilon$, the momentum equation is

$$\upsilon_x \frac{\partial \upsilon_x}{\partial x} + \upsilon_y \frac{\partial \upsilon_x}{\partial y} = \nu \frac{\partial^2 \upsilon_x}{\partial y^2} - \overline{\beta} g (c_i - c_\infty) \sin \epsilon. \tag{27.11}$$

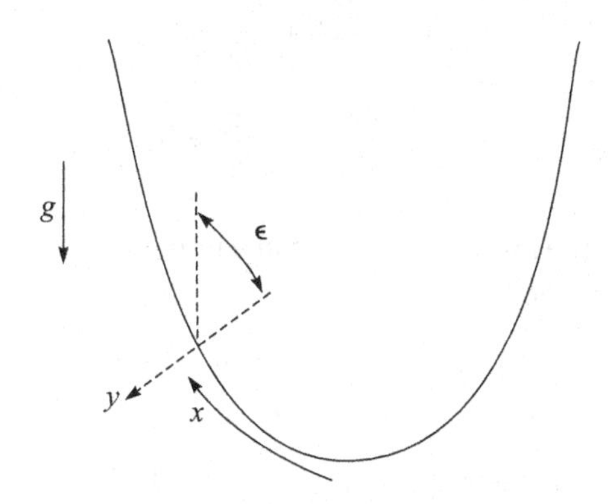

Figure 27.1 Boundary-layer coordinates for free convection.

Boundary conditions for a constant concentration on the surface are

$$\left.\begin{aligned} c_i = c_0, \upsilon_x = \upsilon_y = 0 \quad &\text{at} \quad y = 0. \\ c_i = c_\infty, \upsilon_x = 0 \quad &\text{at} \quad y = \infty. \end{aligned}\right\} \tag{27.12}$$

Solutions to the boundary-layer problem have been obtained for several situations. For example, for a vertical plate ($\sin \epsilon = 1$),

$$\left.\frac{\partial c_i}{\partial y}\right|_{y=0} = \frac{3}{4} C(Sc) \left[\frac{g\bar{\beta}(c_\infty - c_0)}{\nu D x}\right]^{\frac{1}{4}} (c_\infty - c_0), \tag{27.13}$$

where C is a function of the Schmidt number.

Table 27.1 Dependence of the coefficient C on Sc for natural convection at a vertical plate.

Sc	0.01	0.733	1	10	100	1000	∞
C	0.242	0.5176	0.5347	0.6200	0.6532	0.6649	0.6705

Some attention has been devoted to the limit of large Schmidt numbers. Morgan and Warner have shown that within the diffusion layer, the equations can be simplified to [1]

$$\left.\begin{aligned} \nu\frac{\partial^2 v_x}{\partial y^2} - \bar{\beta} g(c_i - c_\infty)\sin\epsilon = 0. \\ \frac{\partial v_x}{\partial x} + \frac{\partial v_y}{\partial y} = 0. \\ v_x\frac{\partial c_i}{\partial x} + v_y\frac{\partial c_i}{\partial y} = D\frac{\partial^2 c_i}{\partial y^2}. \end{aligned}\right\} \tag{27.14}$$

In this region, the inertial terms become unimportant. The boundary conditions become

$$\left.\begin{aligned} c_i = c_0,\ v_x = v_y = 0 \ \text{ at } \ y = 0. \\ c_i = c_\infty,\ \partial v_x/\partial y = 0 \ \text{ at } \ y = \infty. \end{aligned}\right\} \tag{27.15}$$

Outside the diffusion layer, $c_i = c_\infty$, and the velocity decays to zero.

Acrivos has shown that a similarity transformation can be used to reduce this problem to ordinary differential equations for any geometry, that is, any variation of ϵ with x (see Problem 27.1) [2].

The nature of the concentration and velocity profiles and their variation with the Schmidt number can be seen from Fig. 27.2, plotted for a vertical plate from the results of Ostrach [3]. These profiles show, for example, how the diffusion layer becomes thinner at high Schmidt numbers and how the velocity approaches a constant at the outside edge of the diffusion layer and then decays to zero slowly in the region of the hydrodynamic boundary layer outside the diffusion layer.

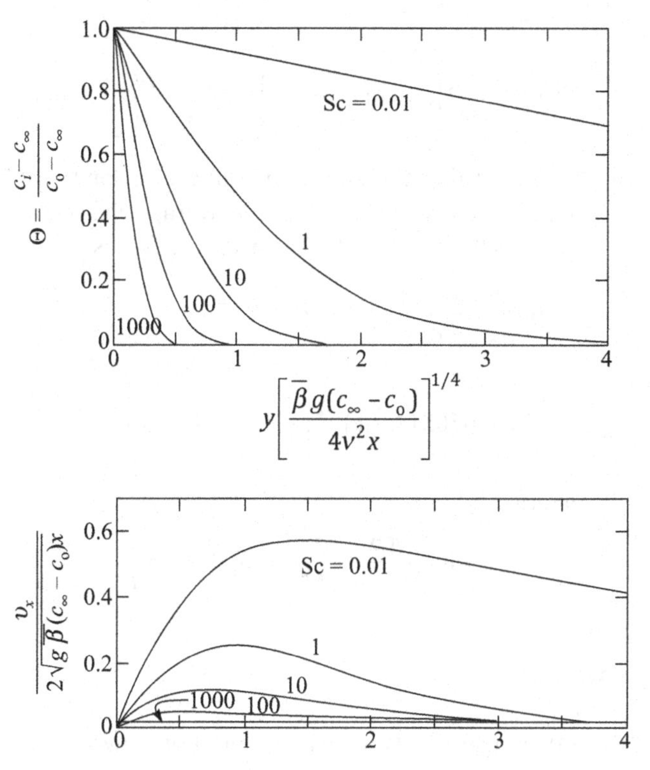

Figure 27.2 Concentration and velocity profiles for free convection at a vertical plate.

Problem

27.1 (a) Show that the boundary-layer equations for free convection at high Schmidt numbers at an arbitrary, two-dimensional surface can be reduced to ordinary differential equations by means of the similarity transformation

$$\Theta = \frac{c_i - c_\infty}{c_o - c_\infty} = \Theta(\eta), \quad \psi = A(x)f(\eta), \quad \eta = y / B(x),$$

where

$$v_x = \frac{\partial \psi}{\partial y}, \; v_y = -\frac{\partial \psi}{\partial x},$$

$$A(x) = \left[\frac{4}{3} D \int_0^x (\sin \epsilon)^{1/3} dx \right]^{3/4} \left[\frac{(c_\infty - c_0)\bar{\beta} g}{\nu} \right]^{1/4},$$

$$B(x) = \left[\frac{4}{3} D \int_0^x (\sin \epsilon)^{1/3} dx \right]^{1/4} \left[\frac{(c_\infty - c_0)\bar{\beta} g}{\nu} \right]^{-1/4} (\sin \epsilon)^{-1/3}.$$

(b) Assuming that the entries in Table 27.1 for the vertical plate are correct, show that the average Nusselt number from $x = 0$ to $x = L$ for the arbitrary surface is

$$\text{Nu} = \frac{1}{c_\infty - c_0} \int_0^L \left. \frac{\partial c_i}{\partial y} \right|_{y=0} dx$$

$$= 0.6705 (\text{Sc}\,\text{Gr})^{1/4} \left[\frac{1}{L} \int_0^L (\sin \epsilon)^{1/3} dx \right]^{3/4},$$

where

$$\text{Gr} = \frac{g(\rho_\infty - \rho_0)L^3}{\rho_\infty \nu^2} \quad \text{and} \quad \text{Sc} = \frac{\nu}{D}.$$

References

1. George W. Morgan and W. H. Warner. "On Heat Transfer in Laminar Boundary Layers at High Prandtl Number." *Journal of the Aeronautical Sciences*, **23**, 937–948 (1956).
2. Andreas Acrivos. "A theoretical analysis of laminar natural convection heat transfer to non-Newtonian fluids." *AIChE Journal*, **6**, 584–590 (1960).
3. Simon Ostrach. "An Analysis of Laminar Free-Convection Flow and Heat Transfer About a Flat Plate Parallel to the Direction of the Generating Body Force." Report 1111. *Thirty-Ninth Annual Report of the National Advisory Committee for Aeronautics, 1953, Including Technical Reports Nos. 1111 to 1157*. Washington: United States Government Printing Office, 1955.

Chapter 28

High Rates of Mass Transfer

One aspect of mass transfer, the interfacial velocity, does not have an analogue in heat transfer. The fact that mass is being transferred through the interface implies that the mass-average velocity is not zero with respect to the interface. The boundary condition at the surface is modified, the velocity profile is distorted, and this can have a considerable effect on the calculated rate of mass transfer [1–4]. This effect vanishes at low rates of mass transfer, and the calculations of preceding chapters must be regarded as valid in the limit where the concentration difference between the surface and the bulk solution is small.

This can be illustrated simply for linear diffusion of A through stagnant B, where the velocity is due solely to the mass-transfer process itself. Such a situation, involving the evaporation of a liquid, is depicted in Fig. 28.1. Here, since the flux of B is zero, the flux of A can be expressed as

$$\mathbf{N}_A = -c\mathcal{D}_{AB}\nabla x_A + c_A\mathbf{v}^* = -c\mathcal{D}_{AB}\nabla x_A + x_A\mathbf{N}_A$$

or

$$\mathbf{N}_A = -\frac{c\mathcal{D}_{AB}\nabla x_A}{1-x_A}. \tag{28.1}$$

At the steady state, $\mathbf{N}_A$ is independent of position, and integration for constant c and $\mathcal{D}_{AB}$ gives

The Newman Lectures on Transport Phenomena
John Newman and Vincent Battaglia

ISBN 978-981-4774-27-7 (Hardcover), 978-1-315-10829-2 (eBook)
www.jennystanford.com

$$N_A = \frac{c\mathcal{D}_{AB}}{L}\ln\frac{1-x_{AL}}{1-x_{A0}}. \tag{28.2}$$

In this system, $N_B = 0$ even though there is a gradient of the concentration of B. The diffusion flux is equal and opposite to the convective flux for species B. For small x_{AL} and x_{A0}, the flux in Eq. 28.2 reduces to $\mathcal{D}_{AB}(c_{A0} - c_{AL})/L$, the same result which would be obtained by the method of preceding chapters with neglect of the velocity due to mass transfer:

$$\mathbf{v}\cdot\nabla c_A \approx 0 = \mathcal{D}_{AB}\nabla^2 c_A.$$

$$N_A = -\mathcal{D}_{AB}\nabla c_A = \mathcal{D}_{AB}\left(c_{A0} - c_{AL}\right)/L.$$

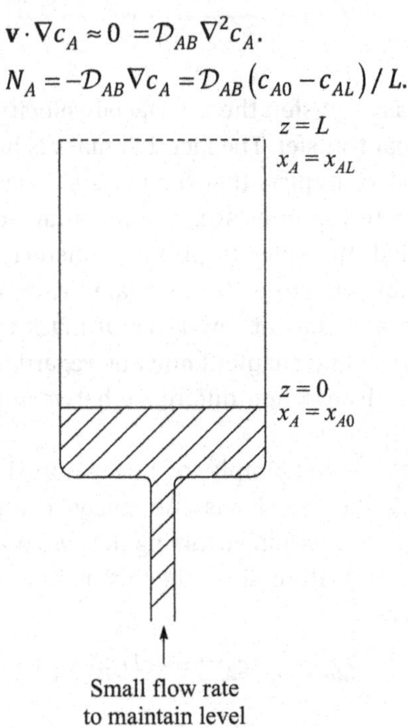

Figure 28.1 Evaporation of liquid A through a gas B.

Similar things happen in more general problems of convection and diffusion. When the hydrodynamic problem becomes coupled through the nonzero interfacial velocity, we should want to use the mass-average velocity, since this is involved in the momentum equation. With the assumption of constant ρ, μ, and $\mathcal{D}_{AB}$, the equations of motion, continuity, and convective diffusion remain (see Chapter 13):

$$\left.\begin{aligned}\rho\left(\frac{\partial \mathbf{v}}{\partial t}+\mathbf{v}\cdot\nabla\mathbf{v}\right)&=-\nabla\mathcal{P}+\mu\nabla^2\mathbf{v}.\\ \nabla\cdot\mathbf{v}&=0.\\ \frac{\partial\omega_A}{\partial t}+\mathbf{v}\cdot\nabla\omega_A&=\mathcal{D}_{AB}\nabla^2\omega_A.\end{aligned}\right\} \tag{28.3}$$

However, the boundary conditions at the surface are modified. Let y be the distance from the surface.

$$\mathbf{n}_A=-\rho\mathcal{D}_{AB}\nabla\omega_A+\rho\omega_A\mathbf{v}=-\rho\mathcal{D}_{AB}\nabla\omega_A+\omega_A(\mathbf{n}_A+\mathbf{n}_B). \tag{28.4}$$

Let $\mathbf{n}_B = 0$ at the interface. Then we have

$$n_{Ay}=-\frac{\rho\mathcal{D}_{AB}}{1-\omega_A}\frac{\partial\omega_A}{\partial y}\bigg|_{y=0}=\rho\upsilon_o, \tag{28.5}$$

where υ_o is the interfacial velocity. The boundary conditions, thus, are written as follows:

$$\text{at } y=0,\ \upsilon_x=0,\ \omega_A=\omega_{A0},\ \upsilon_y=\upsilon_o=-\frac{\mathcal{D}_{AB}}{1-\omega_A}\frac{\partial\omega_A}{\partial y}. \tag{28.6}$$

This assumes that the body shape does not change appreciably during the mass-transfer process.

It is instructive to put the problem in dimensionless form in order to see what additional parameters might be involved. Let

$$\mathbf{v}^*=\frac{\mathbf{v}}{\upsilon_\infty},\quad \mathbf{r}^*=\frac{\mathbf{r}}{L},\quad \Theta=\frac{\omega_A-\omega_{A0}}{\omega_{A\infty}-\omega_{A0}},\quad \mathcal{P}^*=\frac{\mathcal{P}-\mathcal{P}_\infty}{\frac{1}{2}\rho\upsilon_\infty^2}. \tag{28.7}$$

Thus, for a steady state with a uniform surface concentration, the differential equations become

$$\left.\begin{aligned}\mathbf{v}^*\cdot\nabla^*\mathbf{v}^*&=-\frac{1}{2}\nabla^*\mathcal{P}^*+\frac{1}{\text{Re}}\nabla^{*2}\mathbf{v}^*,\\ \nabla^*\cdot\mathbf{v}^*&=0,\\ \mathbf{v}^*\cdot\nabla^*\Theta&=\frac{1}{\text{ReSc}}\nabla^{*2}\Theta,\end{aligned}\right\} \tag{28.8}$$

where Re = $\upsilon_\infty L/\nu$ and Sc = $\nu/\mathcal{D}_{AB}$. These are the same dimensionless groups that have been involved in the mass-transfer problems treated earlier. The boundary conditions at the surface become

$$\text{at } y^* = 0, \; v_x^* = 0, \; \Theta = 0, \; v_y^* = \frac{B}{\text{ReSc}} \frac{\partial \Theta}{\partial y^*}. \tag{28.9}$$

This then introduces a new parameter

$$B = \frac{\omega_{A0} - \omega_{A\infty}}{1 - \omega_{A0}}. \tag{28.10}$$

For $B > 0$, mass transfer is from the surface (blowing), and for $B < 0$, mass transfer is to the surface (suction). We see that $\partial\Theta/\partial y^*$ at $y^* = 0$ will now depend on B as well as geometry and Re and Sc. It so depends on B through the modified velocity near the surface.

The local rate of mass transfer can be expressed in terms of the Nusselt number as

$$\text{Nu}(x) = -\frac{N_y L}{\mathcal{D}_{AB}(c_\infty - c_0)} = \frac{1}{1 - \omega_{A0}} \frac{\partial \Theta}{\partial y^*}\bigg|_{y^*=0} = \frac{\theta \text{Nu}_0(x)}{1 - \omega_{A0}}, \tag{28.11}$$

where $\text{Nu}_0(x)$ is the local Nusselt number as calculated in earlier chapters

$$\text{Nu}_0(x) = \lim_{B \to 0} \frac{\partial \Theta}{\partial y^*}\bigg|_{y^*=0}, \tag{28.12}$$

and θ is a correction factor given by

$$\theta = \frac{\left.\dfrac{\partial \Theta}{\partial y^*}\right|_{y^*=0}}{\text{Nu}_0(x)} = \frac{\left.\dfrac{\partial \Theta}{\partial y^*}\right|_{y^*=0}}{\lim\limits_{B \to 0} \left.\dfrac{\partial \Theta}{\partial y^*}\right|_{y^*=0}}. \tag{28.13}$$

It is convenient to introduce this correction factor because it is not particularly sensitive to the geometry of the problem or to the values of the Reynolds and Schmidt numbers.

As an example, consider the rotating disk. It is still possible to use the von Kármán transformation:

$$\left.\begin{aligned} &\zeta = z\sqrt{\frac{\Omega}{\nu}}, \quad \Theta = \frac{\omega_A - \omega_{A0}}{\omega_{A\infty} - \omega_{A0}}, \\ &H = \frac{v_z}{\sqrt{\nu \Omega}}, \quad G = \frac{v_\theta}{r\Omega}, \quad F = \frac{v_r}{r\Omega}. \end{aligned}\right\} \tag{28.14}$$

The applicable equations then become

$$\left.\begin{aligned} G'' &= HG' + 2FG, \\ F'' &= HF' + F^2 - G^2, \\ 2F + H' &= 0, \\ \Theta'' &= \mathrm{Sc}H\Theta', \end{aligned}\right\} \tag{28.15}$$

as before (see Eqs. 16.9 and 20.2). The boundary conditions now become

$$\left.\begin{array}{l} \text{at } \zeta = 0,\ \ \Theta = 0,\ \ F = 0,\ \ G = 1,\ \ H = B\Theta'/\mathrm{Sc}. \\ \text{at } \zeta = \infty,\ \ \Theta = 1,\ \ F = G = 0. \end{array}\right\} \tag{28.16}$$

Thus υ_z and Θ are still not functions of r, and the mass-transfer rate is still uniform over the surface of the disk and proportional to the square root of the rotation speed.

The resulting correction factors θ for the Nusselt number are shown in Fig. 28.2. For $B < 0$ (suction), the correction factor θ is greater than 1, and conversely. Different curves are obtained for different Schmidt numbers. The curve for Sc = ∞ is given implicitly by

$$\theta = \frac{\Gamma(4/3)}{\displaystyle\int_0^\infty e^{-x^3} \exp\left\{\frac{xB\theta}{\Gamma(4/3)}\right\} dx}. \tag{28.17}$$

This same curve applies for Sc = ∞ to the correction factor for the boundary layer on a flat plate and has been shown by Acrivos to apply to arbitrary, two-dimensional boundary layers at high Schmidt numbers [3]. Results for a flat plate at other values of the Schmidt number are given by Bird, Stewart, and Lightfoot. It is interesting that the curves for finite Schmidt numbers lie on the opposite side of the curve for Sc = ∞ for the flat plate and the rotating disk.

The modification of the velocity profiles by the nonzero rate of mass transfer also has an effect on the fluid friction, the torque in the case of the rotating disk. This effect is shown in Fig. 28.3 in the curve marked θ_υ. Here θ_υ is the correction factor for the torque:

$$\theta_\upsilon = \frac{G'(0)}{\lim\limits_{B\to 0} G'(0)} = -\frac{G'(0)}{0.61589}, \tag{28.18}$$

and this is plotted versus the "rate factor" ϕ_υ, where

$$\phi_\upsilon = -\frac{H(0)}{\lim\limits_{B\to 0} G'(0)} = \frac{H(0)}{0.61589}. \tag{28.19}$$

Figure 28.3 also shows the correction factor for mass transfer for values of the Schmidt number equal to ∞, 1, 0.6, and 0.01, this time plotted against the "rate factor" ϕ for mass transfer:

$$\phi = B\theta. \tag{28.20}$$

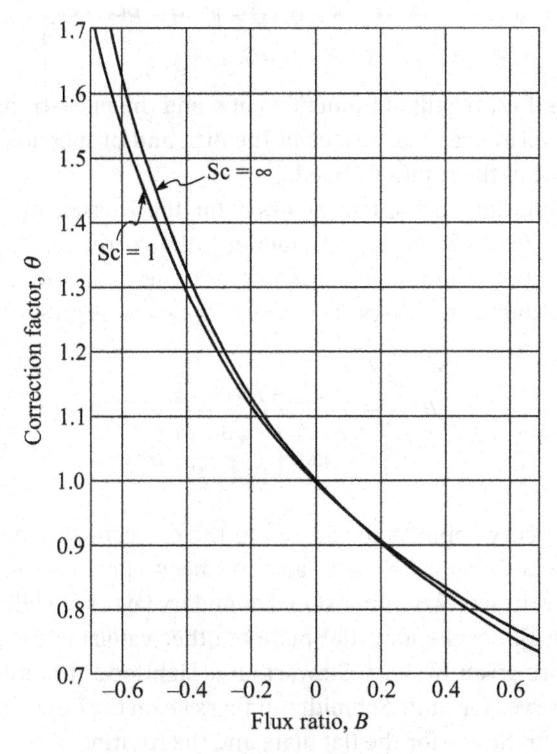

Figure 28.2 The effect of an interfacial velocity on mass-transfer rates for a rotating disk.

To calculate the torque, one first uses B and Sc to get the correction factor θ for the mass-transfer rate, from either Fig. 28.2 or Fig. 28.3. The interfacial velocity $H(0)$ is then related to the rate of mass transfer by Eq. 28.16. With this value of $H(0)$, one can calculate ϕ_υ from Eq. 28.19 and obtain the correction factor θ_υ for the torque from Fig. 28.3.

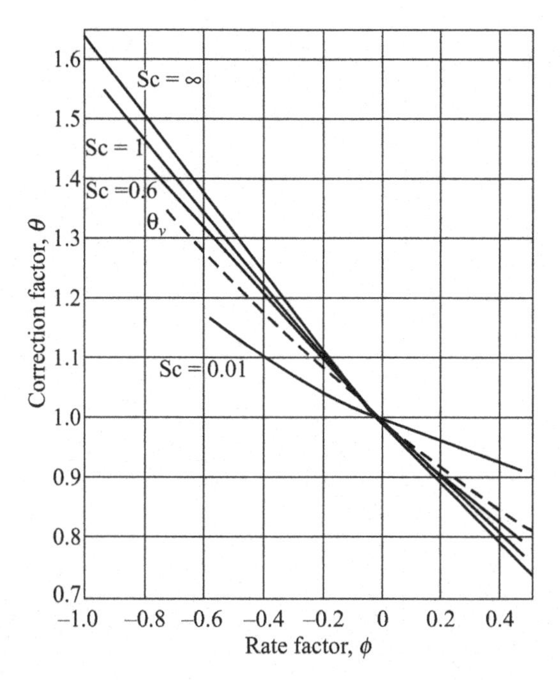

Figure 28.3 The effect of an interfacial velocity on the mass-transfer coefficients and torque for a rotating disk.

References

1. R. Byron Bird, Warren E. Stewart, and Edwin N. Lightfoot. *Transport Phenomena*, 2nd Edition. New York: John Wiley & Sons, 2002, pp. 703–716.
2. Andreas Acrivos. "Mass Transfer in Laminar Boundary-Layer Flows with Finite Interfacial Velocities." *AIChE Journal*, **6**, 410–414 (1960).
3. Andreas Acrivos. "The asymptotic form of the laminar boundary-layer mass-transfer rate for large interfacial velocities." *Journal of Fluid Mechanics*, **12**, 337–357 (1962).
4. Donald R. Olander. "Rotating Disk Flow and Mass Transfer." *Journal of Heat Transfer*, **84C**, 185 (1962).

Chapter 29

Heterogeneous Reaction at a Flat Plate

Consider a first-order, irreversible, heterogeneous reaction at a flat plate in Blasius flow [1]. The equations of motion and continuity are (Chapter 21)

$$v_x \frac{\partial v_x}{\partial x} + v_y \frac{\partial v_x}{\partial y} = \nu \frac{\partial^2 v_x}{\partial y^2}. \tag{29.1}$$

$$\frac{\partial v_x}{\partial x} + \frac{\partial v_y}{\partial y} = 0. \tag{29.2}$$

For high Péclet numbers, the material balance on the reacting species takes the diffusion-layer form (Chapter 24)

$$v_x \frac{\partial c_i}{\partial x} + v_y \frac{\partial c_i}{\partial y} = D \frac{\partial^2 c_i}{\partial y^2}. \tag{29.3}$$

No homogeneous reaction term appears in the material balance because the reaction occurs only on the surface of the plate. Instead, the reaction is treated as a boundary condition

$$D \frac{\partial c_i}{\partial y} = k c_i \quad \text{at} \quad y = 0, \tag{29.4}$$

where k is a rate constant for the reaction. Other boundary conditions on Eq. 29.3 include

$$c_i = c_\infty \text{ at } y = \infty \quad \text{and} \quad c_i = c_\infty \text{ at } x = 0. \tag{29.5}$$

The Newman Lectures on Transport Phenomena
John Newman and Vincent Battaglia

ISBN 978-981-4774-27-7 (Hardcover), 978-1-315-10829-2 (eBook)
www.jennystanford.com

Near the surface, the velocity profile can be approximated by (Chapter 24)

$$\upsilon_x = \beta y, \ \upsilon_y = \frac{a\upsilon_\infty^{3/2} y^2}{4\nu^{1/2} x^{3/2}}, \ \beta = \frac{a\upsilon_\infty^{3/2}}{(\nu x)^{1/2}}. \tag{29.6}$$

This approximation would be valid in Eq. 29.3 for systems where the Schmidt number is large. It is useful to reduce the number of parameters in the problem by nondimensionalizing Eqs. 29.3 through 29.5. First, we can introduce a dimensionless concentration, $\Theta = c_i/c_\infty$, which eliminates c_∞ from the problem. Then Eq. 29.3 becomes,

$$\frac{a\upsilon_\infty^{3/2} y}{(\nu x)^{1/2}} \frac{\partial \Theta}{\partial x} + \frac{a\upsilon_\infty^{3/2} y^2}{4\nu^{1/2} x^{3/2}} \frac{\partial \Theta}{\partial y} = D \frac{\partial^2 \Theta}{\partial y^2}. \tag{29.7}$$

Equation 29.4 shows that k/D can be eliminated by defining a dimensionless normal distance as

$$Y = yk/D. \tag{29.8}$$

Substitution into Eq. 29.7 and rearrangement lead to

$$\frac{a\upsilon_\infty^{3/2} D^2}{\nu^{1/2} k^3} \left[\frac{Y}{x^{1/2}} \frac{\partial \Theta}{\partial x} + \frac{Y^2}{4x^{3/2}} \frac{\partial \Theta}{\partial Y} \right] = \frac{\partial^2 \Theta}{\partial Y^2}. \tag{29.9}$$

Now we can eliminate all of the remaining parameters by defining a dimensionless distance along the plate by

$$X = \frac{xk^2 \nu^{1/3}}{a^{2/3} \upsilon_\infty D^{4/3}} = \frac{xk^2 \mathrm{Sc}^{1/3}}{a^{2/3} \upsilon_\infty D}. \tag{29.10}$$

Now Eq. 29.9 becomes

$$\frac{Y}{X^{1/2}} \frac{\partial \Theta}{\partial X} + \frac{Y^2}{4X^{3/2}} \frac{\partial \Theta}{\partial Y} = \frac{\partial^2 \Theta}{\partial Y^2}. \tag{29.11}$$

The boundary conditions are

$$\frac{\partial \Theta}{\partial Y} = \Theta \text{ at } Y = 0, \quad \Theta = 1 \text{ at } Y = \infty, \quad \Theta = 1 \text{ at } X = 0. \tag{29.12}$$

This is a zero-parameter problem as stated. Because of this generality, it is worthwhile to work out the solution despite the difficulty of obtaining a closed-form solution. The principle of superposition allows us to restate the problem in terms of an integral equation. We can write down a solution for the case in which mass-transfer limitations dominate. This corresponds to replacing the kinetic boundary condition with a unit-step change in the surface concentration

$$\Theta = 0 \quad \text{at} \quad Y = 0 \quad \text{for} \quad X > X'. \tag{29.13}$$

The dimensionless mass transfer to the plate for this case is obtained from the Lighthill transformation (Chapter 24, Problem 24.2, or Eqs. 17.67 and 17.95 in Ref. [2])

$$\left.\frac{\partial \Theta}{\partial Y}\right|_{Y=0} = \frac{1}{X^{1/4}\Gamma(4/3)[12(X^{3/4} - X'^{3/4})]^{1/3}}. \tag{29.14}$$

Let

$$\frac{1}{\Gamma(4/3)12^{1/3}} = 0.489138 = C_2. \tag{29.15}$$

Duhamel's superposition integral allows us to express the solution to the present problem in terms of an integral along the length of the plate of properly weighted solutions for the unit-step change in concentration. In general form, Duhamel's integral is

$$\Theta(X,Y) = 1 - [1 - \Theta(0,0)]B(X,Y;\,0) + \int_0^X \left.\frac{d\Theta}{dX}\right|_{\substack{X=X'\\Y=0}} B(X,Y;X')dX', \tag{29.16}$$

where B is the solution to Eq. 29.11 but with a vanishing value at infinity and a positive unit-step change in the surface value at $X = X'$:

$$B = 0 \text{ at } Y = 0 \quad \text{for} \quad X < X',$$

$$B = 1 \text{ at } Y = 0 \quad \text{for} \quad X \geq X', \tag{29.17}$$

and $$B = 0 \text{ at } X = 0 \quad \text{and} \quad \text{at } Y = \infty.$$

Since we are interested in the flux density to the plate, we can differentiate both sides of Eq. 29.16 with respect to Y and evaluate this at the surface

$$\left.\frac{\partial \Theta}{\partial Y}\right|_{Y=0} = \int_0^X \left.\frac{\partial \Theta}{\partial X}\right|_{\substack{X=X'\\Y=0}} \left.\frac{\partial B}{\partial Y}\right|_{Y=0} dX'. \tag{29.18}$$

Substitution of Eq. 29.14 into Eq. 29.18 gives

$$\left.\frac{\partial\Theta}{\partial Y}\right|_{Y=0} = \frac{-C_2}{X^{1/4}} \int_0^X \left.\frac{\partial\Theta}{\partial X}\right|_{\substack{X=X'\\Y=0}} \frac{dX'}{(X^{3/4} - X'^{3/4})^{1/3}}. \quad (29.19)$$

Further let $Z = X^{3/4}$. After rearrangement

$$\left.\frac{\partial\Theta}{\partial Y}\right|_{Y=0} = \frac{-C_2}{Z^{1/3}} \int_0^Z \left.\frac{\partial\Theta}{\partial Z}\right|_{\substack{Z=Z'\\Y=0}} \frac{dZ'}{(Z - Z')^{1/3}} = \Theta(Z, Y=0). \quad (29.20)$$

By incorporating the boundary condition 29.12, we have obtained an integral equation for $\Theta(Z, Y = 0)$, which is now a complete formulation of the problem.

We found above that we could reduce this problem to an integral equation, but how can we solve the resulting integral? Note that the integral equation is of the Volterra, or initial value, type. This means that the solution for Θ at position Z' involves Θ only at earlier positions ($Z < Z'$), i.e., information does not propagate upstream. These equations are solved by starting at the initial condition ($Z = 0$, where $\Theta = 1$) and marching along the plate, solving for Θ as we march. We solve this here by employing the numerical method of Acrivos and Chambré [3], which involves breaking the integral up into many small steps in Z. Then, taking $Z = (n - l)\Delta z$, we find

$$\left.\frac{\partial\Theta}{\partial Y}\right|_{Y=0} = \frac{-C_2}{Z^{1/3}} \sum_{j=2}^{n} \int_{(j-2)\Delta z}^{(j-1)\Delta z} \left.\frac{\partial\Theta}{\partial Z}\right|_{\substack{Z=Z'\\Y=0}} \frac{dZ'}{[(n-1)\Delta z - Z']^{1/3}}. \quad (29.21)$$

If the size of Δz is small enough, we can assume that $\left.\frac{\partial\Theta}{\partial Z}\right|_{\substack{Z=Z'\\Y=0}}$ is a constant over each step, giving

$$\left.\frac{\partial\Theta}{\partial Y}\right|_{Y=0} = \frac{-C_2}{Z^{1/3}} \sum_{j=2}^{n} \frac{\theta_j - \theta_{j-1}}{\Delta z} \int_{(j-2)\Delta z}^{(j-1)\Delta z} \frac{dZ'}{[(n-1)\Delta z - Z']^{1/3}}. \quad (29.22)$$

These integrals can be evaluated directly, giving

$$\left.\frac{\partial\Theta}{\partial Y}\right|_{Y=0} = \frac{-C_2}{Z^{1/3}} \sum_{j=2}^{n} \frac{\Theta_j - \Theta_{j-1}}{\Delta z} \left(-\frac{3}{2}\right) \times \left([(n-1)\Delta z - (j-1)\Delta z]^{2/3} - [(n-1)\Delta z - (j-2)\Delta z]^{2/3}\right). \quad (29.23)$$

Or, with further manipulation,

$$\left.\frac{\partial\Theta}{\partial Y}\right|_{Y=0} = \frac{1.5C_2}{(Z\Delta z)^{1/3}}\left(\sum_{j=2}^{n}\Theta_j\left[(n-j)^{2/3}-(n-j+1)^{2/3}\right]\right.$$
$$\left.-\sum_{J=1}^{n-1}\Theta_J\left[(n-J-1)^{2/3}-(n-J)^{2/3}\right]\right). \tag{29.24}$$

Or,

$$\left.\frac{\partial\Theta}{\partial Y}\right|_{Y=0} = \frac{1.5C_2}{(Z\Delta z)^{1/3}}\left(-\Theta_n+\Theta_1\left[(n-1)^{2/3}-(n-2)^{2/3}\right]\right.$$
$$\left.+\sum_{j=2}^{n-1}\Theta_j\left[2(n-j)^{2/3}-(n-j+1)^{2/3}-(n-j-1)^{2/3}\right]\right). \tag{29.25}$$

And finally we arrive at

$$\left.\frac{\partial\Theta}{\partial Y}\right|_{Y=0} = \frac{1.5C_2}{(Z\Delta z)^{1/3}}\left(-\Theta_n+\Theta_1 B_{n-1}+\sum_{j=2}^{n-1}\Theta_j A_{n-j}\right)=\Theta_n, \tag{29.26}$$

where for convenience we have defined

$$B_i = i^{2/3}-(i-1)^{2/3}, \tag{29.27}$$

and

$$A_i = 2i^{2/3}-(i+1)^{2/3}-(i-1)^{2/3}. \tag{29.28}$$

Equation 29.26 is for the unknown concentration Θ_n; all prior values of Θ_i, $i < n$, are known. The array of A_i is constant and can be tabulated a priori and used as necessary during the numerical routine. Note that in this problem, we have $\Theta_1 = 1$.

A brief computer program for solving Eq. 29.26 is as follows:

```
c        Program to solve for surface reaction at a flat plate
c        (Blasius flow) with a first-order, irreversible
c        reaction, see Levich [1], section 17.
```

```
      implicit real*8 (a-h,o-z)
      dimension E(201), AA(201),BB(201),th(201),X(201)
      common e,th,nztl,al,be,AA,BB,cl,c2,ex,dz
      Lmax=201
      ex=2.0d0/3.0d0
      do 29 L=1,Lmax
      a=L
      aa(L)=2.0d0*A**ex-(A+1.0d0)**ex-(A-1.0d0)**ex
   29 bb(L)=a**ex-(a-1.0d0)**ex
      dz=0.1d0
      do 1 L=1,Lmax
      z=dz*dble(L-1)
      X(L)=z**(4.0d0/3.0d0)
   1  continue
      ex=4.0d0/3.0d0
      c2=1.11984652/12.0**(1.0d0/3.0d0)
      cl=0.0
      nztl=201
      th(1)=1.0d0
      call theta
      print 101, (x(j),th(j),j=1, Lmax)
  101 format (2x,f10.3,f12.5)
      end
      subroutine theta
      implicit real*8 (a-h,o-z)
c     Subprogram for calculating concentration
c     dated from about December 3, 1967
      dimension E(201), A(201), b(201), th(201)
      common e, th, nztl, al, be, a, b, cl, c2, ex, dz
      nzt=nztl-1
      do 60 nz=2,nztl
      z=dble(nz-1)*dz
      sum=0.0
      if(nz.le.2) go to 42
c     calc. sum(th(j)*a(k))
      do 40 j=3, nz
      k=nz-j+1
```

```
40    sum=sum+th(j-1)*a(k)
42    continue
      nj=nz-1
      th(nz)-(th(1)*b(nj)+sum)/((z*dz)**(1.0d0/3.0d0)/1.5/
      c2+1.0d0)
60    continue
      return
      end
```

This program has been used to determine the reaction rate at the plate as a function of dimensionless distance *X*; a graph of this is given in Fig. 29.1. The asymptote for large *X* represents the mass-transfer-limited solution. Kinetic limitations exist for small values of *X*. The region where the curve (representing the reaction rate) begins to decrease can be regarded as a region where both kinetics and mass transfer are important (mixed mode). The form of *X* in Eq. 29.10 shows that this transition occurs at a physical distance *x* of the order $v_\infty D^{4/3}/k^2 \nu^{1/3}$. The readers can test whether this result follows their intuition.

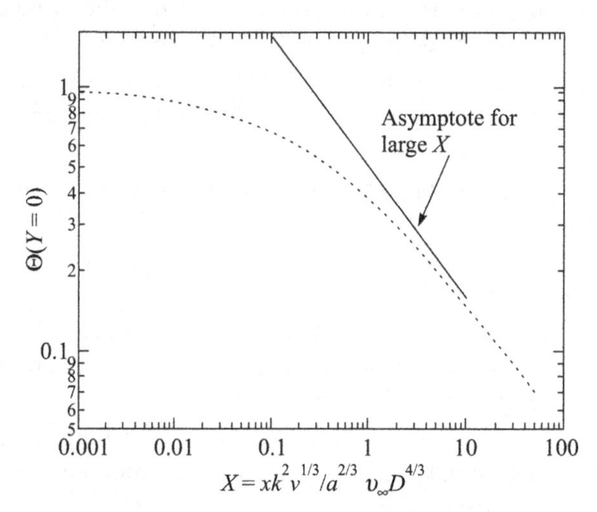

Figure 29.1 Mass transfer to a flat plate at high Schmidt numbers, with a first-order, irreversible surface reaction. Θ(*Y*=0) is proportional to the reaction rate.

Problems

29.1 Blasius flow prevails on a semi-infinite flat plate. A species present in a dilute solution in the free stream and having a very small diffusion coefficient undergoes an irreversible reaction at the surface of the plate. The kinetics of the reaction are found to be nth order in the concentration of the reactant.

(a) What partial differential equation governs transport of the species to the surface? A general name or an equation in vector notation is sufficient here.

(b) Express the heterogeneous reaction kinetics as a mathematical boundary condition on the equation from Part (a).

(c) What simplifications are permissible in the structure of the equation of Part (a) in its application to this problem? State each simplification mathematically, and state its justification in physical terms.

(d) What boundary conditions are appropriate for this equation (in addition to that given in Part (b))?

(e) Eliminate as many parameters from the problem as possible by defining new variables and parameters.

(f) Ascertain whether a similarity solution is possible. Choose a form that appeals to you and discuss whether or not it will work.

29.2 Consider the boundary condition used at the upstream end of the flat plate ($\Theta = 1$).

(a) What is the physical justification of this boundary condition?

(b) Formulate the problem in the case in which axial diffusion is important, including the proper boundary condition at the upstream end of the plate.

(c) If the flat plate considered here were finite in extent, what would the proper boundary condition be at the downstream end?

29.3 Justify the steps taken in going from Eq. 29.16 to Eq. 29.18.

29.4 Apply Problem 24.2 to the present situation in order to derive Eq. 29.14.

29.5 Verify that boundary condition 29.4 correctly represents the physics and chemistry of the problem as stated.

29.6 Show how a in Eq. 29.6 is related to $f''(0)$ for the Blasius solution for flow over a flat plate (see, for example, Fig. 21.2).

References

1. Veniamin G. Levich. *Physicochemical Hydrodynamics*, Englewood Cliffs, N.J.: Prentice-Hall, Inc., 1962.
2. John Newman and Nitash P. Balsara. *Electrochemical Systems*, 4th Edition. Hoboken, New Jersey: John Wiley and Sons, 2020.
3. A. Acrivos and P. L. Chambré, "Laminar Boundary Layer Flows with Surface Reactions," *Industrial & Engineering Chemistry*, **49**, 1025 (1957).

Chapter 30

Mass Transfer to the Rear of a Sphere in Stokes Flow

Boundary-layer methods typically break down at the rear of a bluff object. There are a few cases where this phenomenon can be elucidated. In these cases, the failure of the boundary-layer methods is confined to a small region behind the object, and this region becomes smaller as the Péclet number or the Reynolds number increases. Mass transfer at high Schmidt numbers to a sphere in Stokes flow is an example of such a case [1].

The diffusion-layer treatment of Chapter 19 fails when θ is of order ϵ, where $\epsilon = (2/\text{Pe})^{1/3}$. This is mainly because the diffusion layer becomes thick, and the velocity profiles can no longer be approximated by their forms near the solid surface (see Problem 19.1). However, a more detailed analysis shows the necessity for considering not just one, but four additional regions. Three of these are shown in Fig. 30.1, along with the diffusion layer and the outer region of uniform concentration. The sixth region is in the far wake along the rear axis.

The appropriate stretched coordinates for each region are shown in Table 30.1. Since these variables are of order unity within their region, they also show the size of each region in terms of θ and y. The equation of convective diffusion 19.2 allows different approximations for high Péclet numbers depending on which region is involved. These approximate forms are also shown in the table. We have already worked extensively with the approximations valid in the diffusion layer (region 2).

The Newman Lectures on Transport Phenomena
John Newman and Vincent Battaglia

ISBN 978-981-4774-27-7 (Hardcover), 978-1-315-10829-2 (eBook)
www.jennystanford.com

Table 30.1 Appropriate variables and approximate form of the equation of convection diffusion in different regions for mass transfer to a sphere in Stokes flow.

Region	Variables	Approximate form of equation of convective diffusion	Important modes of mass transfer
1	r^* and θ	$\mathbf{v}^*.\nabla^*\Theta = 0$	Convection
2	θ and $Y = y/\epsilon$	$\frac{3}{2}Y^2\cos\theta\frac{\partial\Theta}{\partial Y} - \frac{3}{2}Y\sin\theta\frac{\partial\Theta}{\partial\theta} = \frac{\partial^2\Theta}{\partial Y^2}$	Convection, radial diffusion
3	r^* and $S = \theta/\epsilon$	$\mathbf{v}^*.\nabla^*\Theta = 0$	Convection
4	r^* and $s = \theta/\epsilon^{3/2}$	$\left(1 - \frac{3}{2r^*} + \frac{1}{2r^{*3}}\right)\frac{\partial\Theta}{\partial r^*} - \left(1 - \frac{3}{4r^*} - \frac{1}{4r^{*3}}\right)\frac{s}{r^*}\frac{\partial\Theta}{\partial s} = \frac{1}{r^{*2}}\left(\frac{\partial^2\Theta}{\partial s^2} + \frac{1}{s}\frac{\partial\Theta}{\partial s}\right)$	Convection, tangential diffusion
5	$Y = y/\epsilon$ and $S = \theta/\epsilon$	$\frac{3}{2}Y^2\frac{\partial\Theta}{\partial Y} - \frac{3}{2}YS\frac{\partial\Theta}{\partial S} = \frac{\partial^2\Theta}{\partial Y^2} + \frac{\partial^2\Theta}{\partial S^2} + \frac{1}{S}\frac{\partial\Theta}{\partial S}$	Convection, radial diffusion, tangential diffusion
6	$\mathcal{R} = \epsilon r^*$ and $\mathcal{S} = \theta/\epsilon^2$	$\frac{\partial\Theta}{\partial\mathcal{R}} - \frac{\mathcal{S}}{\mathcal{R}}\frac{\partial\Theta}{\partial\mathcal{S}} = \frac{1}{\mathcal{R}^2}\left(\frac{\partial^2\Theta}{\partial\mathcal{S}^2} + \frac{1}{\mathcal{S}}\frac{\partial\Theta}{\partial\mathcal{S}}\right)$	Convection, diffusion in θ-direction

$y = r^* - 1$, $\epsilon = (2/\mathrm{Pe})^{1/3}$

A picture of the various regions can be developed by proceeding logically from one region to another, examining the dominant modes of mass transfer in each region and the limits of validity of the treatment for each region, either by observing whether the boundary conditions are satisfied or by examining the order of magnitude of terms neglected in the equation of convective diffusion. The fact that a limit of validity is found indicates the existence of an adjoining region in which different mechanisms of mass transfer are important.

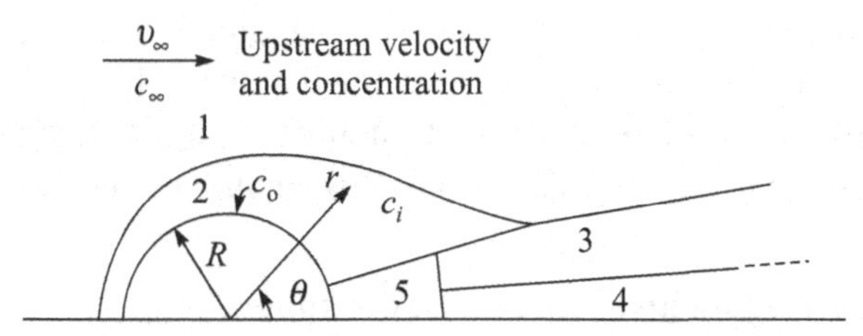

Figure 30.1 Regions of dominance of different mechanisms of mass transfer, for a sphere in Stokes flow at high Péclet numbers.

Sometimes it is possible to infer the existence of an adjoining region without actually obtaining the solution in the first region. For example, the approximate form of the equation of convective diffusion in the first region may indicate that the boundary conditions cannot, in general, be satisfied. On the other hand, the form of the complete equation of convective diffusion may indicate certain neglected terms that should become important in an adjoining region.

Region 1: Region far from the sphere

In this region, the presence of the object of a different concentration is not felt by the fluid, and its concentration is uniform and equal to the upstream concentration, $\Theta = 1$. Consequently, there is no diffusion. Mass is simply carried along by convection.

Region 2: The diffusion (boundary) layer

In this thin region, where $y = O(\epsilon)$, the velocity becomes small, and radial diffusion becomes of the same importance as convection. The solution obtained in Chapter 19 is given in Eqs. 19.9, 19.12, and 19.15.

Region 3: Convective region

As the diffusion layer becomes thick near the rear axis, the velocity is no longer small, and convection again becomes the dominant mode of mass transfer. However, the concentration is not uniform since this region receives the depleted fluid from the diffusion layer. Instead, the concentration is constant along streamlines.

Region 4: The rear-axis region

The solution in the convective region still does not satisfy the symmetry condition $\partial\Theta/\partial\theta = 0$ at $\theta = 0$. This can be traced to the neglect of the tangential diffusion terms in region 3 and indicates the presence of still another region near the rear axis where convection and diffusion in the θ-direction are important, but diffusion in the r-direction is not. At the same time, Θ is small, of order $\sqrt{\epsilon}$ in this region.

Region 5: Region near the rear of the sphere

This region is close to the surface at the rear of the sphere, where the velocity is still small. In this region, both radial and tangential diffusion are of the same importance as convection, and Θ is only of order ϵ. Appropriate variables in this region are $Y = (r^* - 1)/\epsilon = y/\epsilon$ and $S = \theta/\epsilon$, and the equation of convective diffusion reduces in the limit of large Péclet numbers to

$$\frac{3}{2}Y^2\frac{\partial\Theta}{\partial Y} - \frac{3}{2}YS\frac{\partial\Theta}{\partial S} = \frac{\partial^2\Theta}{\partial Y^2} + \frac{\partial^2\Theta}{\partial S^2} + \frac{1}{S}\frac{\partial\Theta}{\partial S}. \quad (30.1)$$

The boundary conditions are

1. $\Theta = 0$ at $Y = 0$.
2. $\partial\Theta/\partial S = 0$ at $S = 0$.
3. $\Theta \to \eta\Theta'\ (\eta \to 0) = \epsilon YS/\Gamma(4/3)(3\pi)^{1/3} =$ as $S \to \infty$, in order to match with the diffusion-layer solution of region 2.
4. For $Y \to \infty$, an asymptotic solution can be found based on the fact that $\partial^2\Theta/\partial Y^2$ must become negligible as regions 3 and 4 are approached.

Since region 5 is in contact with the surface, the normal concentration gradient at the surface will yield the rate of mass transfer in the rear region. Thus, it is essential to obtain the solution in this region. As $Y \to \infty$, the term $\partial^2\Theta/\partial Y^2$ must become negligible, and Θ must satisfy the equation

$$\frac{3}{2}Y^2\frac{\partial\Theta}{\partial Y}-\frac{3}{2}YS\frac{\partial\Theta}{\partial S}=\frac{\partial^2\Theta}{\partial S^2}+\frac{1}{S}\frac{\partial\Theta}{\partial S} \tag{30.2}$$

with the solution

$$\Theta=\epsilon Y^{1/2}f(\xi)\ \text{ where }\ \xi=S\sqrt{Y} \tag{30.3}$$

and f satisfies the ordinary differential equation

$$\xi f''+(1+3\xi^2/4)f'-3\xi f/4=0 \tag{30.4}$$

with the boundary conditions

$$f'=0 \text{ at } \xi=0$$

and

$$f\rightarrow\xi/\Gamma(4/3)(3\pi)^{1/3} \text{ as } \xi\rightarrow\infty.$$

This thus provides a fourth boundary condition for region 5.

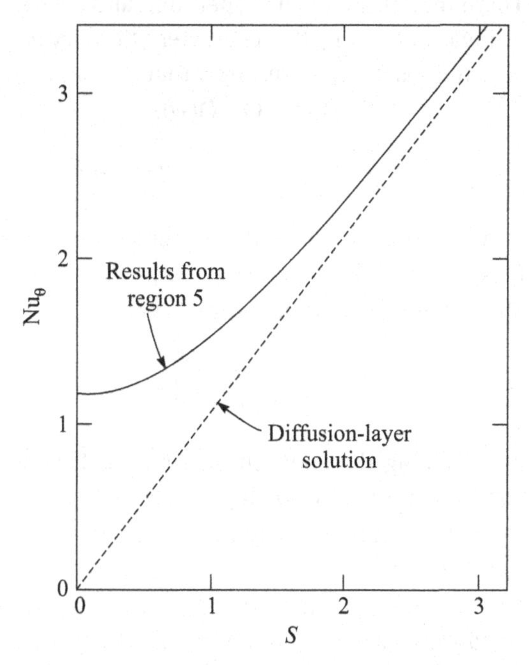

Figure 30.2 Local Nusselt number in the region near the rear of the sphere.

Since Θ and y are both of order ϵ in region 5, the local Nusselt number is of order unity in this region, in contrast to the front part of the sphere where the Nusselt number is of order $1/\epsilon$. Numerical solution to the equation for region 5 yields the local Nusselt number shown in Fig. 30.2. This blends in with the diffusion-layer solution as $S \to \infty$ and has the value 1.192 at the rear of the sphere. This is lower than the value (Nu = 2) for diffusion from a sphere into a stagnant medium because the convection of depleted solution into the rear region decreases the concentration gradient at the rear.

Problems

30.1 For each region, substitute the variables shown in Table 30.1 into the equation of convective diffusion 19.2 and obtain the approximate form valid in the limit of large Péclet numbers.

30.2 (a) Obtain the form of the equation of convective diffusion appropriate to the convective region (region 3).

(b) This equation should be hyperbolic, and the concentration is invariant along the characteristic curves, which are streamlines in this case. Show that

$$\Theta = \Theta(\psi),$$

where $$\psi = \frac{S^2}{3^{5/3}\pi^{2/3}}\left(\frac{1}{r*} - 3r* + 2r*^2\right),$$

is a solution to the diffusion equation derived above. (Actually, it is the general solution.)

(c) Show that the solution in region 3 is

$$\Theta = \frac{1}{\Gamma(4/3)}\int_0^{\sqrt{\psi}} e^{-x^3}\,dx$$

by matching the solution obtained in Part (b) with the solution in the diffusion layer (region 2).

(d) Does this solution satisfy the symmetry condition $\partial\Theta/\partial\theta = 0$ at $\theta = 0$?

(e) Use this solution to provide a boundary condition for region 4. What other boundary conditions are appropriate for region 4?

30.3 Verify the matching condition of region 5 with region 2 (the diffusion layer).

30.4 Consider mass transfer at high Schmidt numbers to a symmetric, bluff cylinder. The velocity profiles near the solid surface are

$$v_x = y\beta(x) \text{ and } v_y = -\frac{1}{2}y^2\beta'(x),$$

and near the rear of the cylinder, β goes to zero like

$$\beta \to A(x_0 - x),$$

where x_0 is the value of x at the rear of the cylinder. For Reynolds numbers that are not too high, there should be no eddies behind the cylinder, and β will be nonnegative.

The Lighthill transformation (see Problem 24.2) provides a solution to the diffusion-layer equation on the front part of the cylinder for high Schmidt numbers. Suggest and defend a method for calculating the rate of mass transfer near the rear of the cylinder. Give a differential equation for the concentration and a complete set of boundary conditions. The problem so posed should be independent of the perturbation parameter.

30.5 Show how the results for mass transfer to the rear of a sphere in Stokes flow can be applied to the rear of an axisymmetric, bluff body at high Schmidt numbers in the absence of eddies behind the body, that is, where the velocity profiles near the solid surface can be approximated by

$$v_x = y\beta(x) \text{ and } v_y = -\frac{1}{2}y^2\frac{(\mathcal{R}\beta)'}{\mathcal{R}},$$

where β is nonnegative and where β goes to zero near the rear of the body like

$$\beta \to A(x_0 - x) \text{ as } x \to x_0 \text{ and } \mathcal{R} \to x_0 - x,$$

where x_0 is the value of x at the rear of the body. The procedure is similar to that for Problem 30.4, but the results of Problem 24.5 should be used for the diffusion layer.

30.6 For the rotating disk and stationary plane in Problem 16.1, the radial and normal velocity distributions near the stationary plane can be expressed as

$$v_r = -Ar\Omega y\sqrt{\Omega/\nu} \text{ and } v_y = A\Omega y^2\sqrt{\Omega/\nu}, \qquad (30.5)$$

where y is the normal distance from the plane and A is a positive constant. For an active section of radius r_0 embedded in the stationary plane, mass transfer in the diffusion layer was treated in Problem 24.7 for high Schmidt numbers.

(a) The diffusion-layer solution would be expected to break down at $r = 0$. Therefore, let us concentrate on the region near $r = 0$ and $y = 0$. Write the equation of convective diffusion for this region, without approximating the diffusion terms. The convective velocity can be approximated by Eq. 30.5. Rewrite the equation of convective diffusion in terms of the variables

$$Y = y\sqrt{\frac{\Omega}{\nu}}\left(\frac{2A\nu}{3D_i}\right)^{1/3} \quad \text{and} \quad S = r\sqrt{\frac{\Omega}{\nu}}\left(\frac{2A\nu}{3D_i}\right)^{1/3}. \tag{30.6}$$

(b) Compare the mass-transfer equation obtained in Part (a) with that solved for region 5 at the rear of a sphere in Stokes flow; see Chapter 30. Is it possible to use the solution presented there to predict the mass-transfer rate near $r = 0$ for the active area on the stationary plane?

30.7 In using the results of Chapters 16 and 20 for the fluid flow and mass transfer to a rotating disk, one is frequently concerned about edge effects. Consider a mass-transfer disk of radius r_o embedded in a larger, inert disk. In this way, one can circumvent edge effects in the hydrodynamics and concentrate on those for the mass transfer.

(a) Write down the equation of convective diffusion in cylindrical coordinates for this situation. Assume steady-state and axial symmetry.

(b) Simplify the velocity components for distances close to the rotating disk, on the basis of having a large Schmidt number. Write down the equation of convective diffusion in its simplified form.

(c) Why is it justified to simplify the velocity components as in Part (b), when the Schmidt number is large?

(d) The concentration at infinity is c_∞ and that on the mass-transfer disk ($r < r_o$) is c_o. (The flux is zero to the inert disk beyond $r = r_o$.) State the solution that applies in the central region of the mass-transfer disk when edge effects are ignored.

(e) Does the solution in Part (d) cease to apply near the edge of the mass transfer disk because

i. Terms, in the equation of convective diffusion, neglected in obtaining that solution can no longer be neglected?

ii. The solution ceases to satisfy the boundary conditions?

(f) If the equation of Part (b) were replaced by its diffusion-layer form, which of the following statements would be true?

i. In the diffusion terms, all derivatives with respect to r can be neglected compared to the second derivative of z.

ii. The convective terms should not be *further* simplified on the basis of the thinness of the diffusion layer.

iii. The mass-transfer problem is now parabolic instead of elliptic.

iv. The effect of the edge of the mass-transfer disk will propagate upstream a small but nonzero amount.

(g) Assume that the region where the edge effect manifests itself is small compared to the size of the mass-transfer disk. Simplify the equation of Part (b) accordingly.

(h) Formulate a mathematical problem appropriate to this edge region. Define stretched independent variables for this region so that the resulting dimensionless partial differential equation has no parameters (like D, Re, Sc). What is the order of the size of this edge region?

(i) Formulate boundary conditions for the partial differential equation of Part (h). These should also be independent of any parameters.

(j) Is the equation of Part (h) now elliptic, parabolic, or hyperbolic?

Reference

1. Ping Huei Sih and John Newman. "Mass Transfer to the Rear of a Sphere in Stokes Flow." *International Journal of Heat and Mass Transfer*, **10**, 1749–1756 (1967).

Chapter 31

Spin Coating

We take up this problem because spin coating has become an important industrial process in the electronics industry, is an area of current research interest, illustrates application of the equations of motion in the "lubrication" approximation, is an application involving hyperbolic equations, and is an area where one may wish to contemplate extensions to nonnewtonian fluids, simultaneous evaporation and film thinning, and application to nonflat surfaces (where previous device fabrication, with masking and etching, has left an uneven surface).

In this application, a liquid is placed on a disk near the axis of rotation. Subsequent rotation causes the fluid to flow radially outward (eventually flying off the edge of the disk, whence it can be recycled) forming a thin, reasonably uniform film. The rate of thinning of the film decreases markedly because of the strong viscous forces in the thin film. During this time, evaporation of a solvent may occur, so that the final film is no longer liquid. Research has shown that the fluid flow and thinning occur at early times and can be treated approximately with the neglect of evaporation, which occurs at later times. We reproduce the classic treatment [1] of the subject, thereby keeping things simple while setting the stage for later contributions dealing with numerous complications. References [2], [3], and [4] build on this early work.

The Newman Lectures on Transport Phenomena
John Newman and Vincent Battaglia

ISBN 978-981-4774-27-7 (Hardcover), 978-1-315-10829-2 (eBook)
www.jennystanford.com

We can start by writing down the equations of motion in cylindrical coordinates:

r-component of the equation of motion

$$\rho\left(\frac{\partial v_r}{\partial t}+v_r\frac{\partial v_r}{\partial r}+\frac{v_\theta}{r}\frac{\partial v_r}{\partial \theta}-\frac{v_\theta^2}{r}+v_z\frac{\partial v_r}{\partial z}\right)$$
$$=-\frac{\partial p}{\partial r}+\mu\left[\frac{\partial}{\partial r}\left(\frac{1}{r}\frac{\partial}{\partial r}(rv_r)\right)+\frac{1}{r^2}\frac{\partial^2 v_r}{\partial \theta^2}-\frac{2}{r^2}\frac{\partial v_\theta}{\partial \theta}+\frac{\partial^2 v_r}{\partial z^2}\right]+\rho g_r. \tag{31.1}$$

θ-component of the equation of motion

$$\rho\left(\frac{\partial v_\theta}{\partial t}+v_r\frac{\partial v_\theta}{\partial r}+\frac{v_\theta}{r}\frac{\partial v_\theta}{\partial \theta}+\frac{v_r v_\theta}{r}+v_z\frac{\partial v_\theta}{\partial z}\right)$$
$$=-\frac{1}{r}\frac{\partial p}{\partial \theta}+\mu\left[\frac{\partial}{\partial r}\left(\frac{1}{r}\frac{\partial}{\partial r}(rv_\theta)\right)+\frac{1}{r^2}\frac{\partial^2 v_\theta}{\partial \theta^2}+\frac{2}{r^2}\frac{\partial v_r}{\partial \theta}+\frac{\partial^2 v_\theta}{\partial z^2}\right]+\rho g_\theta. \tag{31.2}$$

z-component of the equation of motion

$$\rho\left(\frac{\partial v_z}{\partial t}+v_r\frac{\partial v_z}{\partial r}+\frac{v_\theta}{r}\frac{\partial v_z}{\partial \theta}+v_z\frac{\partial v_z}{\partial z}\right)$$
$$=-\frac{\partial p}{\partial z}+\mu\left[\frac{1}{r}\frac{\partial}{\partial r}\left(r\frac{\partial}{\partial r}(v_z)\right)+\frac{1}{r^2}\frac{\partial^2 v_z}{\partial \theta^2}+\frac{\partial^2 v_z}{\partial z^2}\right]+\rho g_z. \tag{31.3}$$

Equation of continuity

$$\frac{\partial \rho}{\partial t}+\frac{1}{r}\frac{\partial}{\partial r}(\rho r vs_r)+\frac{1}{r}\frac{\partial}{\partial \theta}(\rho v_\theta)+\frac{\partial \rho v_z}{\partial z}=0. \tag{31.4}$$

I always marvel at these equations in cylindrical coordinates because of the appearance of mysterious terms in the *r*- and θ-components and the fact that the derivatives with respect to *r* in the viscous terms in the *z*-component appear in a different manner from those in the *r*- and θ-components.

The first approximation introduced for thin films is to assume that the θ-component of the velocity in the film is everywhere equal to the value of the disk surface:

$$v_\theta = r\Omega, \tag{31.5}$$

where Ω is the constant angular velocity of the disk. Equation 31.1 reduces to

$$\rho\Omega^2 r = -\mu \frac{\partial^2 \upsilon_r}{\partial z^2}, \tag{31.6}$$

with a few additional assumptions:

1. The gravitational force is either negligible or directed in the z-direction.
2. The flow is axisymmetric.
3. Because the film is thin, variations are more rapid in the z-direction than in the r-direction.
4. Because surface forces are small and the film is adjacent to a constant-pressure gas, the pressure variation is negligible.
5. Acceleration in the radial direction is negligible compared to the centrifugal force.

Integration with respect to z gives a parabolic velocity profile

$$\upsilon_r = -\frac{\rho\Omega^2 r}{2\mu}\left(z^2 - 2hz\right), \tag{31.7}$$

after applying the boundary conditions that υ_r is zero at the surface (z = 0) and the shear stress is zero at the free surface (that is, $\partial \upsilon_r / \partial z = 0$ at $z = h$, where $h(r, t)$ is the film thickness.

Integration of the equation of continuity 31.4 with respect to z gives

$$\upsilon_r(z-h) = -\int_0^h \frac{1}{r}\frac{\partial}{\partial r}(r\upsilon_r)dz. \tag{31.8}$$

The left side of this equation is the time derivative of the film thickness, and the right side can be evaluated with the help of Eq. 31.7:

$$\begin{aligned}\frac{\partial h}{\partial t} &= -\frac{1}{r}\frac{d}{dr}\left(r\int_0^h v_r dz\right)\\ &= -\frac{1}{r}\frac{d}{dr}\left(r\frac{\rho\Omega^2 r}{2\mu}\left[-\frac{z^3}{3} + hz^2\right]_0^h\right)\\ &= -\frac{1}{r}\frac{d}{dr}\left(\frac{\rho\Omega^2}{2\mu}r^2h^3\frac{2}{3}\right) \qquad = -\frac{\rho\Omega^2}{3\mu}\frac{1}{r}\frac{d}{dr}(r^2h^3).\end{aligned} \tag{31.9}$$

If we define the constant K as

$$K = \frac{\rho\Omega^2}{3\mu}, \tag{31.10}$$

then the equation reads

$$\frac{\partial h}{\partial t} = -\frac{K}{r}\frac{\partial}{\partial r}(r^2 h^3). \tag{31.11}$$

This constitutes a single partial differential equation for the film thickness h as it varies as a function of r and t.

Equation 31.11 is a hyperbolic equation for the film thickness. We should seek a solution by the method of characteristics [5]. Let us change the variables from the set of r and t to the set of β and t, where β is a characteristic variable, yet to be defined. For this transformation, we need to write

$$\left(\frac{\partial h}{\partial t}\right)_\beta = \left(\frac{\partial h}{\partial t}\right)_r + \left(\frac{\partial h}{\partial r}\right)_t \left(\frac{\partial r}{\partial t}\right)_\beta. \tag{31.12}$$

Substitution of Eq. 31.11 gives

$$\left(\frac{\partial h}{\partial t}\right)_\beta = -2Kh^3 - 3Krh^2\left(\frac{\partial h}{\partial r}\right)_t + \left(\frac{\partial h}{\partial r}\right)_t \left(\frac{\partial r}{\partial t}\right)_\beta. \tag{31.13}$$

Now, let β be defined so that the last two terms in Eq. 31.13 cancel, that is,

$$\left(\frac{\partial r}{\partial t}\right)_\beta = 3Krh^2. \tag{31.14}$$

This equation defines the characteristic curve traced in the $r - t$ plane by a line of constant β. At the same time, Eq. 31.13 simplifies to

$$\left(\frac{\partial h}{\partial t}\right)_\beta = -2Kh^3. \tag{31.15}$$

Thus, Eq. 31.15 shows how the film thickness changes with t along a curve of constant β, and Eq. 31.14 shows how r is varying at the same time. This approach, by the method of characteristics, can be useful even when the resulting problem formulation needs to be solved numerically. This is so because of the simple structure of the equations.

In the present case, an analytic solution can be obtained. Equation 31.15 looks a lot like an ordinary differential equation for h and can be integrated by the following steps:

$$\frac{dh}{h^3} = -2K\,dt\,. \tag{31.16}$$

$$\frac{h^{-2}}{-2} = -2Kt + f(\beta)\,. \tag{31.17}$$

We need to recognize here that the "constant" of integration can be a function of β. Rearrangement gives

$$h^{-2} = 4Kt + h_o^{-2}(\beta), \tag{31.18}$$

or

$$h = \frac{h_o(\beta)}{[1 + 4Kt\,h_o^2(\beta)]^{1/2}}, \tag{31.19}$$

where h_o is the initial thickness of the film. Equation 31.14 now becomes

$$\left(\frac{\partial r}{\partial t}\right)_\beta = 3Kr\frac{h_o^2(\beta)}{[1 + 4Kt\,h_o^2(\beta)]}. \tag{31.20}$$

Since β is being held constant, it is not too difficult to integrate this equation, following the steps:

$$\ln\frac{r}{r_o(\beta)} = \frac{3}{4}\ln[1 + 4Kt\,h_o^2(\beta)] \tag{31.21}$$

or

$$r = r_o(1 + 4Kh_o^2 t)^{3/4}. \tag{31.22}$$

With these two equations 31.19 and 31.22, we now have a description of how h changes with t at a given β and how r also changes with t at a given β. Now it is necessary to contemplate how to use these equations with a given initial (axisymmetric) distribution of liquid on the disk.

Problem

31.1 In a spin-coating problem, the initial film thickness distribution is $h_o = a/(1 + \alpha^2 r_o^2)^{1/4}$, where r_o is the radial position and a and α are constants. On a given disk, the thickness initially varies by a factor of 5. For $Ka^2t = 9.5$, calculate the film thickness at the center and the edge of the disk. Here $K = \rho\Omega^2/3\mu$, where ρ and μ are the density and viscosity of the fluid and Ω is the angular velocity of the disk.

References

1. Alfred G. Emslie, Francis T. Bonner, and Leslie G. Peck, "Flow of a Viscous Liquid on a Rotating Disk," *J. Appl. Phys.*, **29**, 858–862 (1958).
2. Taku Ohara, Yoichiro Matsumoto, and Hideo Ohashi, "The film formation dynamics in spin coating," *Phys. Fluids*, **A1**, 1949–1959 (1989).
3. David E. Bornside, "Mechanism for the Local Planarization of Microscopically Rough Surfaces by Drying Thin Films of Spin-Coated Polymer/Solvent Solutions," *J. Electrochem. Soc.*, **137**, 2589–2595 (1990).
4. L. M. Peurrung and D. B. Graves, "Film Thickness Profiles over Topography in Spin Coating," *J. Electrochem. Soc.*, **138**, 2115–2124 (1991).
5. Andreas Acrivos, "Method of Characteristics Technique. Application to Heat and Mass Transfer Problems," *Ind. Eng. Chem.*, **48**, 703–710 (1956).

Chapter 32

Stefan–Maxwell Mass Transport

Mathematical solutions for multicomponent mass transfer in complex situations are not very plentiful. It is desirable to have some idealized problems that illustrate physical and chemical phenomena even though they do not represent industrial processes in detail. Many such problems are examples of the film model, the penetration model, or the boundary-layer model [1]. We develop here Stefan–Maxwell transport for these models. In this formulation, driving forces for diffusion are expressed as linear combinations of flux densities. For substitution into a material balance, it might be attractive to use an inverse formulation whereby flux densities are expressed as linear combinations of driving forces. However, the transport properties so defined are likely to be much more complex functions of composition than the Stefan–Maxwell coefficients $\mathcal{D}_{ij}^{SM}$. This is particularly true for low-pressure gas mixtures where the coefficients are independent of composition and equal to the values for binary pairs of species, measurable in the absence of a third component and able to be estimated for simple molecules by kinetic theory. It is more fruitful to develop examples where we have a clearer route for obtaining transport properties than to develop elaborate and abstract theories and computational methods where such properties are unlikely to be available. Later we expound on

The Newman Lectures on Transport Phenomena
John Newman and Vincent Battaglia

ISBN 978-981-4774-27-7 (Hardcover), 978-1-315-10829-2 (eBook)
www.jennystanford.com

the wide range of problems of chemistry and geometry that can be covered.

For ideal mixtures, the Stefan–Maxwell equation takes the form

$$\nabla x_i = \sum_j \frac{x_i N_j - x_j N_i}{c\mathcal{D}_{ij}^{\mathrm{SM}}}, \tag{32.1}$$

where x_i is the mole fraction and N_i is the flux density of species i, c is the total solution concentration, and $\mathcal{D}_{ij}^{\mathrm{SM}}$ is the diffusion coefficient for the i–j pair of species. $\mathcal{D}_{ji}^{\mathrm{SM}} = \mathcal{D}_{ij}^{\mathrm{SM}}$ by the Onsager reciprocal relations. Generalizations are available for nonideal phases and for situations with pressure diffusion or thermal diffusion [2–7]. For ideal-gas mixtures, the superscript SM becomes superfluous, and $c\mathcal{D}_{ij}$ is independent of pressure as well as composition.

Film model: From the literature, we choose the example of the evaporation of a liquid mixture of acetone and methanol through a stagnant layer of air [8] (treated as a single component) and of thickness δ. Since momentum considerations are not very important, the pressure is essentially constant, as well as the temperature. This means that c and $\mathcal{D}_{ij}$ are also constants. It is convenient to define a dimensionless flux density $\mathcal{N}_i$ as

$$\mathcal{N}_i = \frac{N_{iz}\delta}{c\mathcal{D}_{10}}, \tag{32.2}$$

where $\mathcal{D}_{10}$ is a convenient diffusion coefficient. (When we make dilute-solution approximations, we envision species 0 as the solvent.) A dimensionless distance η can be defined by

$$\eta = z/\delta. \tag{32.3}$$

For the steady-state film model, the material balance equations (one for each species) become

$$\frac{d\mathcal{N}_i}{d\eta} = 0. \tag{32.4}$$

There are only $n - 1$ independent Stefan–Maxwell equations for a solution with n species. In dimensionless form, Eq. 32.1 becomes

$$\frac{dx_i}{d\eta} = \sum_j \frac{x_i \mathcal{N}_j - x_j \mathcal{N}_i}{\mathcal{D}_{ij} / \mathcal{D}_{10}}. \tag{32.5}$$

The last equation is that the mole fractions sum to one:

$$\sum_i x_i = 1 . \tag{32.6}$$

For our system of acetone, methanol, and air, there are three flux densities $\mathcal{N}_i$ and three mole fractions, for a total of six unknowns. Three Eq. 32.4, two Eq. 32.5, and one Eq. 32.6 apply. The five first-order differential equations require five boundary conditions. For our evaporation problem, we state them as

$$x_1 = x_{10} = 0.319,\ x_2 = x_{20} = 0.528,\ \mathcal{N}_0 = 0 \text{ at } \eta = 0, \tag{32.7}$$

$$x_1 = x_{1\delta} = 0,\ \ x_2 = x_{2\delta} = 0 \ \text{ at } \eta = 1 . \tag{32.8}$$

This system can be solved analytically by those with perseverance and algebraic skill. We prefer to emphasize a numerical approach because it is so easy to do and so easy to extend to any number of species and so easy to generalize to problems of greater complexity (see below). Because this is a boundary-value problem, the BAND approach is recommended [7, 9]. The composition profiles in Fig. 32.1 are essentially a reproduction of results in Ref. [8]. These profiles are bowed as a result of the nonzero interfacial velocity due to the evaporation itself. Parameters used in this example are given in Table 32.1.

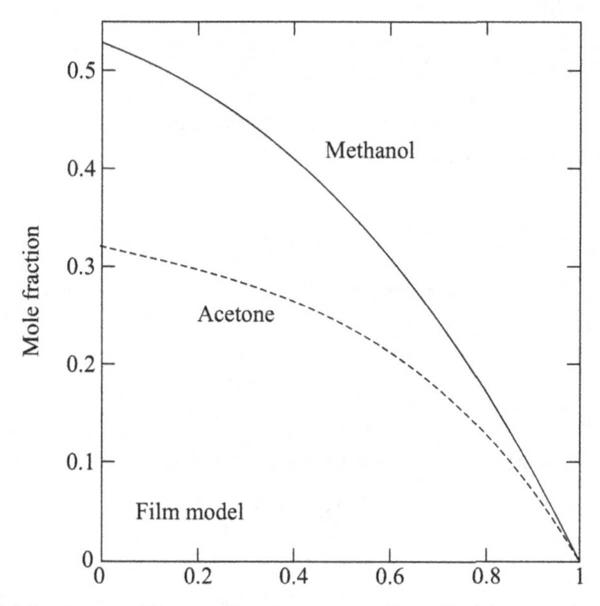

Figure 32.1 Composition profiles for evaporation of acetone and methanol through a stagnant film of air.

Table 32.1 Parameters used in examples of evaporation of acetone and methanol into air.

	1	2	0
	Acetone	Methanol	Air
M_i (g/mol)	58.08	32.04	29
x_{i0}	0.319	0.528	0.153
$x_{i\delta}$ or $x_{i\infty}$	0	0	1
$\mathcal{D}_{10}, \mathcal{D}_{20}, \mathcal{D}_{12}$ (cm^2/s)	0.1372	0.1991	0.0848
T (K), p (bar)	328.5	0.9935	

Penetration model: Many similar problems can be addressed in a transient mode, whereby a similarity transformation can reduce the problem to ordinary differential equations. For example, we can consider the evaporation of acetone and methanol into an infinite vertical column initially filled with air. The material balances take the form (still with constant pressure, temperature, and total concentration c)

$$c\frac{\partial x_i}{\partial t} = -\nabla \cdot N_i . \tag{32.9}$$

The Stefan–Maxwell equations 32.1 remain valid.

The similarity transform variable is

$$\eta = \frac{z}{2\sqrt{\mathcal{D}_{10}t}} . \tag{32.10}$$

The flux densities transform to functions of η according to the definition

$$\mathcal{N}_i(\eta) = \frac{2N_{iz}}{c}\sqrt{\frac{t}{\mathcal{D}_{10}}} . \tag{32.11}$$

After transformation, the material balance equation 32.9 becomes

$$2\eta\frac{dx_i}{d\eta} = \frac{d\mathcal{N}_i}{d\eta} . \tag{32.12}$$

The transformed Stefan–Maxwell equations take the form of Eq. 32.5, and Eq. 32.6 remains applicable. Thus, the problem formulation remains nearly identical to the film model. The boundary conditions can be written as

$$x_1 = x_{10},\ \ x_2 = x_{20},\ \ \mathcal{N}_0 = 0 \ \text{ at } \ \eta = 0, \tag{32.13}$$

$$x_1 \to x_{1\infty},\ \ x_2 \to x_{2\infty} \ \text{ as } \ \eta \to \infty, \tag{32.14}$$

almost like those for the film model. Included here is the collapse of the initial condition at $t = 0$ with the boundary condition for $z \to \infty$.

The composition profiles in Fig. 32.2 are generated by a computer program nearly identical to that used for the film model, with due regard for the difference in the material balance 33.12 and by carrying the calculations to $\eta = \eta_{max} = 6$ to approximate $\eta \to \infty$. The profiles are still bowed near $\eta = 0$ due to the nonzero interfacial velocity, representing the evaporating acetone and methanol. At large η, the profiles flatten out to approach the boundary condition at infinity.

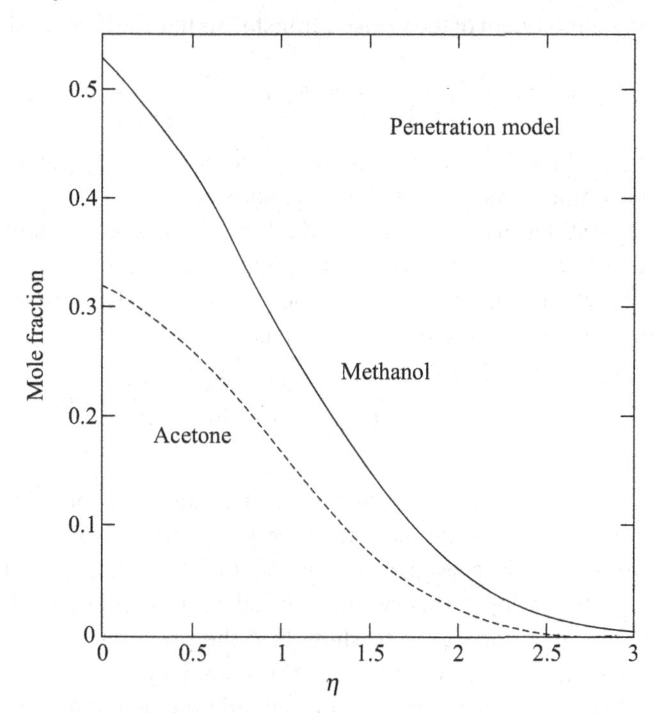

Figure 32.2 Composition profiles for transient evaporation of acetone and methanol into air.

Boundary-layer model: Next we can address the evaporation of acetone and methanol from a flat plate into a flowing air stream. This is a physically realistic situation, and it should be possible to avoid assumptions like having both constant c and ρ, which requires that all the molar masses be the same. Some new parameters become involved here, particularly the molar masses M_i, and the viscosity.

The traditional similarity variable for the flat plate is used here:

$$\eta = y\sqrt{\frac{\upsilon_\infty}{\nu_\infty x}}, \tag{32.15}$$

but with care to use the constant ν_∞ rather than a variable kinematic viscosity to make the problem dimensionless. The essential feature of the transformation is the dependence on x (the distance parallel to the plate) and y (the distance perpendicular to the plate). υ_∞ and ν_∞ are the velocity and kinematic viscosity far from the flat plate. The x-component of the velocity transforms like the mole fractions,

$$V_x(\eta) = \upsilon_x / \upsilon_\infty, \tag{32.16}$$

and the y-component of the velocity transforms like the flux densities,

$$V_y(\eta) = \upsilon_y\sqrt{\frac{x}{\nu_\infty \upsilon_\infty}} \text{ and } \mathcal{N}_i(\eta) = \frac{N_{iy}}{c}\sqrt{\frac{x}{\nu_\infty \upsilon_\infty}}, \tag{32.17}$$

where V_x, V_y, and $\mathcal{N}_i$ represent dimensionless velocity and flux density components in the directions indicated.

But first let us develop the fluid mechanics in a rigorous manner, or at least one in which the approximations become clear. For a Newtonian fluid of variable density ρ and viscosity μ, the x-component of the equation of motion is

$$\rho\left(\frac{\partial \upsilon_x}{\partial t} + \upsilon_x\frac{\partial \upsilon_x}{\partial x} + \upsilon_y\frac{\partial \upsilon_x}{\partial y}\right) = -\frac{\partial p}{\partial x} - \frac{\partial \tau_{xx}}{\partial x} - \frac{\partial \tau_{yx}}{\partial y} + \rho g_x. \tag{32.18}$$

For a fluid of variable density, it is not generally possible to incorporate the gravitational term into a gradient of dynamic pressure $\mathcal{P}$; therefore, we neglect the gravitational term explicitly

In boundary-layer theory, the normal or y-component of the equation of motion is used to show that the pressure variation is negligible through the thickness of the boundary layer. For a flat plate at zero incidence, the pressure outside the boundary layer is uniform, and hence p becomes a constant in this problem. This is in harmony with a constant value of c, used in the film and penetration models. In boundary-layer theory, derivatives with respect to x are small relative to those with respect to y. This means that $\tau_{xx} << \tau_{yx}$ and that the derivatives in Eq. 32.18 doubly insure that the τ_{xx} term is negligible. For a steady state, the equation of motion becomes

$$\rho\left(\upsilon_x\frac{\partial \upsilon_x}{\partial x}+\upsilon_y\frac{\partial \upsilon_x}{\partial y}\right)=-\frac{\partial \tau_{yx}}{\partial y}, \qquad (32.19)$$

a form that has merit even when we get to turbulent boundary layers.

For a Newtonian fluid, the stress relation is

$$\tau_{yx}=-\mu\left(\frac{\partial \upsilon_x}{\partial y}+\frac{\partial \upsilon_y}{\partial x}\right)\approx-\mu\frac{\partial \upsilon_x}{\partial y}, \qquad (32.20)$$

where the boundary-layer approximation again ensures that the derivative with respect to x is negligible. For our problem, the velocity is now governed by the equation

$$\rho\left(\upsilon_x\frac{\partial \upsilon_x}{\partial x}+\upsilon_y\frac{\partial \upsilon_x}{\partial y}\right)=\frac{\partial}{\partial y}\left(\mu\frac{\partial \upsilon_x}{\partial y}\right), \qquad (32.21)$$

and after transformation, we have

$$\frac{\rho}{\rho_\infty}\left(-\frac{\eta}{2}V_x+V_y\right)\frac{dV_x}{d\eta}=\frac{d}{d\eta}\left(\frac{\mu}{\mu_\infty}\frac{dV_x}{d\eta}\right). \qquad (32.22)$$

The x-component of the Stefan–Maxwell equation reads

$$\frac{\partial x_i}{\partial x}=\sum_j\frac{x_iN_{jx}-x_jN_{ix}}{c\mathcal{D}_{ij}}. \qquad (32.23)$$

From inspection of the transformation and consideration of the boundary-layer approximations, we conclude that the term on the left is negligible. Equivalently stated, diffusion in the streamwise or x-direction is negligible in boundary-layer theory. The solution for the flux densities in the x-direction then follows from $x_iN_{jx} = x_jN_{ix}$ as

$$N_{ix} = cx_i\upsilon_x. \qquad (32.24)$$

With this result, the material balance 32.9 in the steady state transforms to

$$\frac{\eta}{2}\frac{dx_iV_x}{d\eta}=\frac{d\mathcal{N}_i}{d\eta}, \qquad (32.25)$$

where we recall that $\mathcal{N}_i$ is the dimensionless y-component of the flux density.

The normal component of the Stefan–Maxwell equation transforms to

$$\frac{dx_i}{d\eta} = \sum_j \frac{x_i \mathcal{N}_j - x_j \mathcal{N}_i}{\mathcal{D}_{ij} / \nu_\infty}, \tag{32.26}$$

which differs from Eq. 32.5 simply as a consequence of the nondimensionalization. Now that fluid mechanics has been introduced, it is appropriate to use ν_∞ in forming η and thereby introduce the Schmidt numbers $Sc_{ij} = \nu_\infty/\mathcal{D}_{ij}$.

There are now eight principal unknowns (for the evaporation example), three $\mathcal{N}_i$, three x_i, V_x, and V_y. The mathematical problem is similar to its former structure, with three material balances 33.25, two Stefan–Maxwell relations 33.26, and one Eq. 32.6, but now supplemented with the x-component of the equation of motion in the form 33.22 and with the definition of υ_y in terms of flux densities,

$$\rho V_y = c\sum_i M_i \mathcal{N}_i. \tag{32.27}$$

There is no need for an overall continuity equation; it is equivalent to a sum over Eq. 32.25. In addition, the density can be given by

$$\rho = c\sum_i M_i x_i, \tag{32.28}$$

and a suitable form needs to be developed for the dependence of μ on the mole fractions. The boundary conditions stay in the form of Eqs. 32.13 and 32.14, augmented by the conditions on V_x:

$$V_x = 0 \text{ at } \eta = 0 \text{ and } V_x \to 1 \text{ as } \eta \to \infty. \tag{32.29}$$

Included in the boundary conditions is the collapse of the condition for $x = 0$ with the condition for $y \to \infty$.

The computer program requires more extensive revision to accommodate the boundary-layer model than the penetration model, and there are additional parameters in the molar masses and the fitting of the viscosity. The calculation domain was extended to $\eta = \eta_{max} = 12$, but the operation of the computer program is substantially the same.

Pseudo-binary mixture

To provide a basis for understanding and correlating the Stefan–Maxwell results and the effect of a nonzero interfacial velocity, it is useful to recall the approximations valid when the solutes (1 and 2) are everywhere dilute in the solvent (species 0). Under this circumstance, each solute is effectively diffusing independently through the solvent. The pseudo-binary mixture where all the diffusion coefficients are equal is another useful approximation. By developing the pseudo-binary mixture first, we can then treat a third component as a minor species in a predominantly binary mixture where the interfacial velocity need not be zero.

If all the diffusion coefficients are equal, $\mathcal{D}_{ij} = \mathcal{D}_{10}$, the species become practically identical, and Eq. 32.5 becomes

$$\frac{dx_i}{d\eta} = x_i \sum_j \mathcal{N}_j - \mathcal{N}_i \sum_j x_j. \tag{32.30}$$

Film model: Here we can write

$$\mathcal{N}_i = -\frac{dx_i}{d\eta} + x_i K \tag{32.31}$$

where

$$K = \sum_j \mathcal{N}_{j0} \tag{32.32}$$

can be regarded as a dimensionless interfacial velocity and $\mathcal{N}_{j0}$ denotes the dimensionless flux density at the interface. In view of Eq. 32.4, $\mathcal{N}_i$ is a constant, and Eq. 32.31 has the solution

$$\Theta_i = \frac{x_i - x_{i0}}{x_{i\delta} - x_{i0}} = \frac{e^{K\eta} - 1}{e^K - 1}, \tag{32.33}$$

satisfying the boundary conditions 33.7 and 33.8. Then

$$\mathcal{N}_{i0} = K\frac{x_{i0}e^K - x_{i\delta}}{e^K - 1} = (x_{i0} - x_{i\delta})H_1(K) + x_{i0}K, \tag{32.34}$$

where

$$H_1(K) = \frac{K}{e^K - 1}. \tag{32.35}$$

Note that, for $K >> 1$, there is strong blowing from $\eta = 0$, and $\mathcal{N}_{i0} \to Kx_{i0}$. For $K << -1$, there is strong blowing toward $\eta = 0$, and $\mathcal{N}_{i0} \to -Kx_{i\delta}$.

Penetration model: Adding Eq. 32.12 over i shows, in view of Eq. 32.6, that the sum of $\mathcal{N}_i$ is independent of η. Hence, Eq. 32.31 still applies with K defined by Eq. 32.32. Equation 32.12 becomes

$$\frac{d^2x_i}{d\eta^2}+(2\eta-K)\frac{dx_i}{d\eta}=0 \tag{32.36}$$

with the solution

$$\Theta_i=\frac{x_i-x_{i0}}{x_{i\infty}-x_{i0}}=1-\frac{\operatorname{erfc}(\eta-K/2)}{\operatorname{erfc}(-K/2)} \tag{32.37}$$

satisfying boundary conditions 32.13 and 32.14. Then

$$\mathcal{N}_{i0}=\frac{2}{\sqrt{\pi}}(x_{i0}-x_{i\infty})H_2(K)+x_{i0}K, \tag{32.38}$$

where

$$H_2(K)=\frac{\exp(-K^2/4)}{\operatorname{erfc}(-K/2)}=\frac{\sqrt{\pi}/2}{\int_0^{\infty}e^{-\eta^2+K\eta}d\eta}. \tag{32.39}$$

Boundary-layer model: We also take all the molar masses to be equal, $M_i = M_j$, so that the density is constant, and we take $\mu = \mu_\infty$. Equation 32.25 can then be summed over i and simplified with Eqs. 32.6 and 32.27 to yield

$$\frac{dV_y}{d\eta}=\frac{\eta}{2}\frac{dV_x}{d\eta}=\frac{1}{2}\frac{d\eta V_x}{d\eta}-\frac{1}{2}V_x. \tag{32.40}$$

If we let f be defined by

$$V_x=\frac{df}{d\eta}, \tag{32.41}$$

then Eq. 32.40 can be integrated to yield

$$V_y=\frac{\eta}{2}V_x-\frac{1}{2}f, \tag{32.42}$$

so that the boundary condition for f becomes $f = -2K$ at $\eta = 0$.

With $M_i = M_j$ and the above definitions, the equation of motion 32.22 becomes

$$2\frac{d^3 f}{d\eta^3} + f\frac{d^2 f}{d\eta^2} = 0 \tag{32.43}$$

subject to the boundary conditions

$$f = -2K, \quad \frac{df}{d\eta} = 0 \text{ at } \eta = 0, \quad \frac{df}{d\eta} \to 1 \text{ as } \eta \to \infty. \tag{32.44}$$

With $K = 0$, this is the statement of the classical Blasius solution for flow past a flat plate. This formulation for the hydrodynamics holds when $M_i = M_j$ and $\mu = \mu_\infty$, independent of the values of the diffusion coefficients $\mathcal{D}_{ij}$.

For the solutes, the Stefan–Maxwell equation 32.26 becomes

$$\mathcal{N}_i = -\frac{\mathcal{D}_{10}}{\nu_\infty}\frac{dx_i}{d\eta} + x_i V_y, \tag{32.45}$$

and the material balance equation 32.25 becomes

$$2\frac{\mathcal{D}_{10}}{\nu_\infty}\frac{d^2 x_i}{d\eta^2} + f\frac{dx_i}{d\eta} = 0 \tag{32.46}$$

with the boundary conditions 32.13 and 32.14.

Following Bird et al. [1] (but with some differences in notation), we let $\Pi(\eta, \Lambda, K)$ be the solution to the problem

$$2\frac{d^2\Pi}{d\eta^2} + \Lambda f\frac{d\Pi}{d\eta} = 0 \tag{32.47}$$

with the boundary conditions

$$\Pi = 0 \text{ at } \eta = 0 \quad \text{and} \quad \Pi \to 1 \text{ as } \eta \to \infty \tag{32.48}$$

and where f is determined by Eq. 32.43 with boundary conditions 32.44. Then

$$x_i = x_{i0} + (x_{i\infty} - x_{i0})\Pi(\eta, \mathrm{Sc}_{10}, K), \tag{32.49}$$

$$\mathcal{N}_{i0} = \frac{\mathcal{D}_{i0}}{\nu_\infty}(x_{i0} - x_{i\infty})\Pi'(0, \mathrm{Sc}_{10}, K) + x_{i0}K, \tag{32.50}$$

and

$$df/d\eta = \Pi(\eta, 1, K).$$

Dilute-solution theory

Suppose that i is a dilute component in a solution that is predominantly composed of species 0 and 1, for which we have already applied the concepts of pseudo-binary mixtures to obtain an idea of the dimensionless interfacial velocity K. This can be the basis of a regular-perturbation expansion in which the concentration level of species i constitutes the perturbation parameter. Thus, any variable, such as K, can be expanded in terms of the base value K^0 and a deviation value K':

$$K \approx K^0 + K'. \tag{32.51}$$

The base values are obtained as in the preceding section by ignoring the presence of the minor components, and these could be designated by the superscript 0 and applied to the major components 0 and 1. The correction terms, including the minor components, would then be governed by a linear problem. However, because we already have numerical means of obtaining the results, we do not want to expend a lot of effort getting complicated results whose validity is already restricted to dilute solutions. Consequently, we make an important simplification that $\mathcal{D}_{i0}$ is the same as $\mathcal{D}_{i1}$, and we designate this quantity as $\mathcal{D}_{im}$. This permits one to select an approximate or average value for $\mathcal{D}_{im}$ in the likely case that $\mathcal{D}_{i0} \neq \mathcal{D}_{i1}$. Heat transfer can also be treated like a trace component in problems of simultaneous heat and mass transfer.

Film model: For the major components, 0 and 1, Eqs. 32.31 through 32.34 apply directly, and it matters little whether a superscript 0 is applied to designate the base case. The requirement that $\mathcal{N}_0 = 0$ in Eq. 32.7 can also be relaxed in favor of a specification of K, and K can be regarded as an independent variable.

For a minor component i, with $\mathcal{D}_{i0} = \mathcal{D}_{i1} = \mathcal{D}_{im}$, the Stefan–Maxwell equation 32.5 becomes

$$\mathcal{N}_i = -\frac{\mathcal{D}_{im}}{\mathcal{D}_{10}}\frac{dx_i}{d\eta} + x_i K, \tag{32.52}$$

where linearization approximations have been made. Since $\mathcal{N}_i$ is still constant, the solution for x_i can be written in the form of Eq. 32.33 but with K replaced by $\mathcal{K}$ where

$$\mathcal{K} = \frac{\mathcal{D}_{10}}{\mathcal{D}_{im}}K. \tag{32.53}$$

Consequently, Eq. 32.34 is replaced by

$$\mathcal{N}_{i0} = \frac{\mathcal{D}_{\text{im}}}{\mathcal{D}_{10}}(x_{i0} - x_{i\delta})H_1(\mathcal{K}) + x_{i0}K \qquad (32.54)$$

for dilute solutes. The same limits stated below Eq. 32.35 apply for large, positive or negative, values of K.

For the major species, there are correction terms that make the perturbation consistent with Eq. 32.6, but these are developed only for the penetration model.

Penetration model: Equations 32.36 through 32.38 apply directly to the major components 0 and 1. For a minor component, the Stefan–Maxwell equation reduces to Eq. 32.52, and the material balance becomes

$$\frac{\mathcal{D}_{im}}{\mathcal{D}_{10}}\frac{d^2x_i}{d\eta^2} + (2\eta - K)\frac{dx_i}{d\eta} = 0. \qquad (32.55)$$

The solution should now be written like Eq. 32.37 but with η replaced by $\eta\sqrt{\mathcal{D}_{10} / \mathcal{D}_{im}}$ and K replaced by $\mathcal{K}$ where

$$\mathcal{K} = \left(\frac{\mathcal{D}_{10}}{\mathcal{D}_{im}}\right)^{1/2} K. \qquad (32.56)$$

The dimensionless flux density is given by

$$\mathcal{N}_{i0} = 2\sqrt{\frac{\mathcal{D}_{im}}{\pi\mathcal{D}_{10}}}(x_{i0} - x_{i\infty})H_2(\mathcal{K}) + x_{i0}K. \qquad (32.57)$$

For the corrections for the major components, it is necessary to linearize the governing equations. After using Eqs. 32.6 and 32.32, the Stefan–Maxwell equation for component 1 can be reduced to

$$\mathcal{N}_1' = -\frac{dx_1'}{d\eta} + x_1'K^0 + x_1^0K' + \left(\frac{\mathcal{D}_{10}}{\mathcal{D}_{im}} - 1\right)\left(x_1^0\mathcal{N}_i' - x_i'\mathcal{N}_1^0\right), \qquad (32.58)$$

and substitution into the material balance gives an equation to solve for the correction x_1':

$$\frac{d^2x_1'}{d\eta^2} + (2\eta - K^0)\frac{dx_1'}{d\eta} = K'\frac{dx_1^0}{d\eta} + \left(\frac{\mathcal{D}_{10}}{\mathcal{D}_{im}} - 1\right)\frac{d}{d\eta}(x_1^0\mathcal{N}_i' - x_i'\mathcal{N}_1^0) \qquad (32.59)$$

subject to the conditions that x_1' is specified at both $\eta = 0$ and as $\eta \to \infty$. This problem could be solved analytically, but it is probably not worth the trouble. Additional terms can be included if there are more than one dilute solute. Note that this reduces to the pseudo-binary mixture if $\mathcal{D}_{im} = \mathcal{D}_{10}$. A similar perturbation expansion can be found in supporting electrolyte theory [7].

Boundary-layer model: If the molar masses are all equal and $\mu = \mu_\infty$, then the hydrodynamics follows Eqs. 32.40 through 32.44. The major components 0 and 1 follow Eqs. 32.45 through 32.50.

The minor solutes, in the regular-perturbation expansion, also satisfy Eqs. 32.45 and 32.46 but with $\mathcal{D}_{10}$ replaced by $\mathcal{D}_{im}$. Thus, the solution can be represented as

$$x_i = x_{i0} + (x_{i\infty} - x_{i0})\Pi(\eta, \mathrm{Sc}_{im}, K), \qquad (32.60)$$

and the flux density at the interface is

$$\mathcal{N}_{i0} = \frac{\mathcal{D}_{im}}{v_\infty}(x_{i0} - x_{i\infty})\Pi'(0, \mathrm{Sc}_{im}, K) + x_{i0}K\,. \qquad (32.61)$$

The same problem applies to the minor components and to the major components, although it is not consistent except under the conditions already discussed. Solving for f and Π can provide a basis for comparison with the results calculated correctly with the full Stefan–Maxwell model.

There is not an analytic solution for Π, although numerical results are summarized by Bird et al. [1]. For high Schmidt numbers, the diffusion layer is very thin compared to the hydrodynamic boundary layer. Then f can be approximated by the first two nonzero terms in a Taylor expansion about $\eta = 0$:

$$f \approx -2K + \frac{1}{2}f''(0)\eta^2\,. \qquad (32.62)$$

If K is also small, as can be expected for solutions with large Sc, then $f''(0) = 0.33206$, the Blasius result for $K = 0$. The concentration profile can be expressed as

$$\Pi = \frac{\int_0^\xi e^{-\xi^3 + K\xi}\,d\xi}{\int_0^\infty e^{-\xi^3 + K\xi}\,d\xi}, \qquad (32.63)$$

where

$$\xi = \left(\frac{\Lambda f''(0)}{12}\right)^{1/3} \eta \text{ and } \mathcal{K} = \left(\frac{12\Lambda^2}{f''(0)}\right)^{1/3} K. \qquad (32.64)$$

Consequently, the derivative can be expressed as

$$\Pi'(0,\Lambda,K) = 0.3387\Lambda^{1/3} H_3(\mathcal{K}) \; (\Lambda \gg 1), \qquad (32.65)$$

where $0.3387 = \left[f''(0)/12\right]^{1/3} / \Gamma(4/3)$ and

$$H_3(K) = \frac{\Gamma(4/3)}{\int_0^\infty e^{-\eta^3 + K\eta} d\eta}. \qquad (32.66)$$

This has some similarity to the second expression in Eq. 32.39 for the penetration model. The flux density for species i at the surface becomes, in the limit of high Schmidt numbers,

$$\mathcal{N}_{i0} = 0.3387\left(\frac{\mathcal{D}_{im}}{\nu_\infty}\right)^{2/3} (x_{i0} - x_{i\infty}) H_3(\mathcal{K}) + x_{i0} K. \qquad (32.67)$$

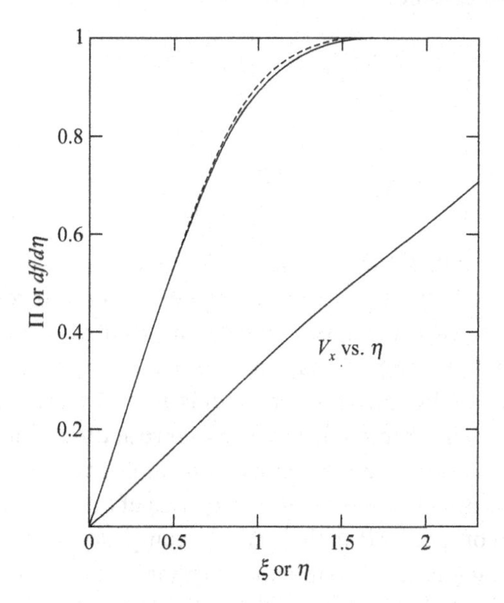

Figure 32.3 Composition profiles in the pseudo-binary case with negligible interfacial velocity. The profile for Sc = 1 (solid curves) is also the velocity profile V_x and is plotted against both ξ and η. The profiles nearly coincide for Sc = 10, 100, 1000, and ∞.

Numerical solutions for $\Pi(\eta, \Lambda, K)$ have been obtained by various researchers. Some profiles for $K = 0$ are given in Fig. 32.3. At $K = 0$, the expressions 33.65 and 33.67 give fair approximations, but the coefficient 0.3387 is actually a weak function of Sc_{im}, having a value of 0.33206 at $Sc_{im} = 1$ and 0.3387 at $Sc_{im} = \infty$ and being within 2% of 0.332 for $0.6 < Sc_{im} \leq \infty$.

Boundary conditions

The example of evaporation of acetone and methanol entails specification of the composition at both boundaries of the fluid. Note that it also requires a condition on one flux density; the absence of such a requirement leaves the system incompletely specified.

An alternative to the composition specification at a boundary would be the statement of all the flux densities. One variant of this is to have a single heterogeneous reaction whereby

$$N_{i0} = s_i r, \tag{32.68}$$

r being the net rate of the process. Let species R be a reactant for which the stoichiometric coefficient s_R is nonzero. Then

$$N_{j0} = \frac{s_j}{s_R} N_{R0}, \tag{32.69}$$

and

$$\sum_j N_{j0} = \frac{N_{R0}}{s_R} \sum_j s_j = \frac{s}{s_R} N_{R0} = sr, \tag{32.70}$$

where s is the sum of the stoichiometric coefficients. The rate r itself might be specified, or the rate might be specified through a rate equation whereby r depends on the composition x_{i0} at the surface.

The reaction rate r may not transform properly for the penetration and boundary-layer models unless there is a limiting reactant whose mole fraction is driven to zero at the surface. Then the boundary conditions would comprise one mole fraction, say x_{R0}, and relationships 32.69 giving the flux densities of other components in terms of that of species R. In this way, the composition of each species at the interface is fixed, implicitly, so that it transforms properly with respect to t or x, in the penetration and boundary-layer models, respectively. The restriction of having a limiting reactant is not necessary for the film model, and it can be relaxed for the boundary layer on a rotating disk or for a two-dimensional or axisymmetric stagnation-point flow.

Range of Problems

A remarkable variety of problems is amenable to analysis by one or more of the models discussed above. Because the composition variables x_i depend on the scaled distance η, problems in which physical properties like ρ, μ, and $\mathcal{D}_{ij}^{\mathrm{SM}}$ depend on composition work well with the similarity transformations for the penetration model and the boundary-layer model, and one can see that this also applies to multicomponent mass transfer according to the Stefan–Maxwell protocol.

Heterogeneous and homogeneous reactions can be handled in the film model and for boundary layers on uniformly accessible surfaces (rotating disk and stagnation-point flows) but may not transform as required for the penetration model and the boundary layer on a flat plate. Exceptions might include an irreversible first-order homogeneous reaction and a very rapid homogeneous reaction, where two limiting reactants are each absent from one region and there is a reaction surface formed between the two regions.

There are important generalization of the models. The quasi-potential transformation permits the film model to be applied to an arbitrary geometry while accommodating an arbitrary number of equilibrated homogeneous chemical reactions, one heterogeneous reaction, variable physical properties, and Stefan–Maxwell transport. Each problem can be separated into a chemical part, which is identical to the standard film model, and a geometric part, in which Laplace's equation is solved for the quasi-potential in the system geometry and with boundary conditions derived from the original problem.

A simple example would be evaporation of acetone and methanol from a sphere of radius R (with neglect of free convection due to buoyancy and with an inexhaustible supply of liquid). Let the flux density of species be related to the quasi-potential η (ordinarily we would call this ϕ, but here η fits conveniently with our notation for the similarity variable) according to

$$N_i = c\mathcal{D}_{10}\mathcal{N}_i(\eta)\nabla\eta\,, \tag{32.71}$$

where η satisfies Laplace's equation

$$\nabla^2\eta = 0 \tag{32.72}$$

with the boundary conditions

$$\eta = 0 \quad \text{at} \quad r = R \quad \text{and} \quad \eta \to 1 \quad \text{as} \quad r \to \infty. \tag{32.73}$$

The steady-state material balance then reduces to Eq. 32.4 since $\nabla^2\eta = 0$ and $(\nabla\eta)^2 \neq 0$. Further postulate that the mole fractions depend only on η, so that the Stefan–Maxwell equations reduce to Eq. 32.5. Since Eq. 32.6 and boundary conditions 32.7 and 32.8 apply, the problem becomes identical to that defined earlier for the film model. The solution to Eq. 32.72 subject to conditions 32.73 is

$$\eta = 1 - \frac{R}{r}. \tag{32.74}$$

Hence, the composition profiles in Fig. 32.1 can be applied to the steady evaporation of acetone and methanol in the spherical geometry and with this interpretation of η in terms of r.

The penetration model can be dated back to Cottrell [10] (and Sand [11]). Application to a growing drop, as it might be formed in solvent extraction or spray drying or in polarography, came with Ilkovič [12–14]. With a growing drop, there is not only the passage of time, as there is with a falling film, but there is also a stretching or elongational flow of the interface and the neighboring fluid. There is a more general transformation for an arbitrary growth rate of the drop [15] and a specialized form when the volumetric growth rate is a constant in time. It is assumed that the diffusion layer is thin compared to the radius of the drop. However, the transformation works without this restriction for the growth of a second phase in a spherical form from a supersaturated system because then the thickness of the diffusion layer and the radius of the drop are both proportional to the square root of time [16]. It is not necessary for physical properties to be constant. Levich did a number of problems with an interface between two fluids, as in mass transfer to a bubble in Hadamard–Rybczyński flow [17]. It was recognized that these penetration-model problems could be generalized to a transformation for mobile interfaces [18, 19].

The Lighthill transformation is an important generalization for boundary-layer problems, although it is restricted to cases where the diffusion layer is thin compared to the hydrodynamic boundary layer, as applies at high Schmidt numbers and for a diffusion layer beginning to grow in a developed hydrodynamic boundary layer. The Lévêque solution for the thermal entrance region in a pipe is an early example, and Levich treated several situations that predate the Lighthill transformation (such as mass transfer to a sphere in Stokes flow) and came up with a statement on the general importance of flows with large Schmidt numbers, important because of the wide significance of aqueous solutions where the Schmidt number approximates 1000. Acrivos showed that variable physical properties and nonzero interfacial velocity could be handled in the context of the Lighthill transformation [20], and Newman [21] applied the method in the context of electrolytic mass transfer.

Liquids could have a negligible dependence of μ, ρ, etc. on p.

If a correlation of $d\Theta_i / d\eta |_{\eta=0}$ versus $\mathcal{D}$ and K were available, it might be possible to resolve the problem. For example, with the evaporation of acetone and methanol, we might have the mole fractions but need to estimate K to get an answer. Bear in mind that it is very easy to solve the problem right with the Stefan–Maxwell model, but it takes a simple computer program. It is also easy to modify that computer program to handle a modified boundary condition such as a reaction rate, which depends on mole fractions or a single limiting reactant.

References

1. R. Byron Bird, Warren E. Stewart, and Edwin N. Lightfoot, *Transport Phenomena*. New York: John Wiley and Sons, 1960.
2. Lars Onsagcr, "Theories and Problems of Liquid Diffusion," *Annals of the New York Academy of Sciences*, **46**, 241–265, 1945.
3. Richard W. Laity, "General Approach to the Study of Electrical Conductance and Its Relation to Mass Transport Phenomena," *Journal of Chemical Physics*, **30**, 682–691, 1959.
4. Richard W. Laity, "An Application of Irreversible Thermodynamics to the Study of Diffusion," *Journal of Physical Chemistry*, **63**, 80–83, 1959.
5. E. N. Lightfoot, E. L. Cussler, Jr., and R. L. Rettig, "Applicability of the Stefan–Maxwell Equations to Multicomponent Diffusion in Liquids," *AIChE Journal*, **8**, 708–710, 1962.

6. John Newman, Douglas Bennion, and Charles W. Tobias, "Mass Transfer in Concentrated Binary Electrolytes," *Berichte der Bunsengesellschaft für physikalische Chemie*, **69**, 608–612, 1965. [For corrections see ibid., **70**, 493, 1966].
7. John Newman and Nitash P. Balsara, *Electrochemical Systems*. Hoboken, New Jersey: John Wiley and Sons, 2020, pp. 236–237 or Appendix C.
8. R. Carty and T. Schrodt, "Concentration Profiles in Ternary Gaseous Diffusion," *Industrial and Engineering Chemistry Fundamentals*, **14**, 276–278, 1975.
9. John Newman, "Numerical Solution of Coupled, Ordinary Differential Equations," *Industrial and Engineering Chemistry Fundamentals*, **7**, 514–517, 1968.
10. F. G. Cottrell, "Der Reststrom bei galvanischer Polarization, betrachet ais ein Diffusuinsproblem," *Zeitschrift fur physicalisch Chemie*, **42**, 385, 1903.
11. Henry J. S. Sand, "III. On the Concentration at the Electrodes in a Solution, with special reference to the Liberation of Hydrogen by Electrolysis of a Mixture of Copper Sulphate and Sulphuric Acid," *The London, Edinburgh, and Dublin Philosophical Magazine and Journal of Science*, Series **6**, 1(1), 45–79, 2010.
12. D. Ilkovič, "Polarographic Studies with the Dropping Mercury Kathode—Part XLIV—The Dependence of Limiting Currents on the Diffusion Constant, on the Rate of Dropping and on the Size of Drops," *Collection of Czechoslovak Chemical Communications*, **6**, 498–513, 1934.
13. D. Ilkovič, "Sur la valcur des courants de diffusion observés dans l'électrolyse a l'aide de 1'électrode a gouttes des mercure: Étude polarographique," *Journal de Chimie Physique*, **35**, 129–135, 1938.
14. D. Mac Gillavry and E. K. Rideal, "On the Theory of Limiting Currents. I. Polarographic limiting currents," *Recueil des Travaux Chimiques des Pays-Bas*, **56**, 1013–1021, 1937.
15. John Newman, "The Fundamental Principles of Current Distribution and Mass Transport in Electrochemical Cells," Allen J. Bard, ed., *Electroanalytical Chemistry*. New York: Marcel Dekker, Inc., 1973, **6**, pp. 187–352.

16. John Bomben, M.S., Department of Chemical Engineering, University of California, Berkeley, ca. 1964.
17. Veniamin G. Levich, *Physicochemical Hydrodynamics*. Englewood Cliffs, New Jersey: Prentice Hall, Inc., 1962, Section 70.
18. J. B. Angelo, E. N. Lightfoot, and D. W. Howard, "Generalization of the Penetration Theory for Surface Stretch: Application to Forming and Oscillating Drops," *AIChE Journal*, **12**, 751–759, 1966.
19. W. E. Stewart, J. B. Angelo, and E. N. Lightfoot, "Forced Convection in Three-Dimensional Flows: II. Asymptotic Solutions for Mobile Interfaces," *AIChE Journal*, **16**, 771–786, 1970.
20. Andreas Acrivos, "Solution of the laminar boundary layer energy equation at high Prandtl numbers," *The Physics of Fluids*, **3**, 657–658, 1960.
21. J. Newman, "The effect of migration in laminar diffusion layers," *International Journal of Heat and Mass Transfer*, **10**, 983–997, 1967.

Suggested Reading for Laminar Flow Solutions

1. Pipe flow, Couette flow, Rayleigh problem.

   Hermann Schlichting, *Boundary-Layer Theory*, 7th Edition, translated by J. Kestin. New York: McGraw-Hill Book Company, 1979, pp. 11–12, 70–82.

   R. Byron Bird, Warren E. Stewart, and Edwin N. Lightfoot, *Transport Phenomena*, 2nd Edition. New York: John Wiley & Sons, 2002 (BSL). pp. 48–53, 114–121, and 142–143.

2. Flow to a rotating disk.

   Hermann Schlichting, *Boundary-Layer Theory*, 7th Edition, translated by J. Kestin. New York: McGraw-Hill Book Company, 1979, pp. 102–107.

   Veniamin G. Levich, *Physicochemical Hydrodynamics*. Englewood Cliffs, N. J.: Prentice-Hall, Inc., 1962, pp. 60–65.

3. Introduction to singular perturbation expansions.

   J. R. Bowen, A. Acrivos, and A. K. Oppenheim, "Singular perturbation refinement to quasi-steady state approximation in chemical kinetics." *Chemical Engineering Science*, **18**, 177, 1963.

   Milton van Dyke, *Perturbation Methods in Fluid Mechanics*. New York/London: Academic Press, 1964.

4. Creeping flow past a sphere.

   Hermann Schlichting, *Boundary-Layer Theory*, 7th Edition, translated by J. Kestin. New York: McGraw-Hill Book Company, 1979, pp. 113–116.

   Veniamin G. Levich, *Physicochemical Hydrodynamics*. Englewood Cliffs, N. J.: Prentice-Hall, Inc., 1962, pp. 395–402,

   BSL, 2nd Edition, pp. 58–61.

   Ian Proudman and J. R. A. Pearson, "Expansions at small Reynolds numbers for the flow past a sphere and a circular cylinder." *Journal of Fluid Mechanics*, **2**, 237–262 (1957).

   Saul Kaplun and P. A. Lagerstrom, "Asymptotic Expansions of Navier–Stokes Solutions for Small Reynolds Numbers." *Journal of Mathematics and Mechanics*, **6**, 585–593 (1957).

5. Mass transfer to a sphere in Stokes flow.

   Andreas Acrivos and Thomas D. Taylor, "Heat and Mass

Transfer from Single Spheres in Stokes Flow." *The Physics of Fluids,* **5**, 387–394 (1962).

Veniamin G. Levich, *Physicochemical Hydrodynamics.* Englewood Cliffs, N. J.: Prentice-Hall, Inc., 1962, pp. 78–87.

Andreas Acrivos and J. D. Goddard, "Asymptotic expansions for laminar forced convection heat and mass transfer. Part 1. Low speed flows." *Journal of Fluid Mechanics,* **23**, 273–291 (1965).

6. Mass transfer to a rotating disk.

   Veniamin G. Levich, *Physicochemical Hydrodynamics.* Englewood Cliffs, N. J.: Prentice-Hall, Inc., 1962, pp. 65–78.

   E. M. Sparrow and J. L. Gregg, "Heat Transfer from a Rotating Disk to Fluids of Any Prandtl Number," *Journal of Heat Transfer,* **81C**, 249–251 (1959).

7. Boundary-layer solutions.

   A. Blasius solution.

      Hermann Schlichting, *Boundary-Layer Theory,* 7th Edition, translated by J. Kestin. New York: McGraw-Hill Book Company, 1979, pp. 127–141.

      BSL, 2nd Edition, pp. 133–140.

   B. Role of coordinate systems.

      Saul Kaplun, "The Role of Coordinate Systems in Boundary-Layer Theory," *Zeitschrift für Angewandte Mathematik und Physik,* **5**, 111–135 (1959).

   C. Higher approximations.

      Milton van Dyke, "Higher Approximations in Boundary-Layer Theory. Part I - General Analysis," Journal of Fluid Mechanics, **14**, 161–177 (1962).

      Milton van Dyke, "Higher Approximations in Boundary-Layer Theory. Part II - Application to Leading Edges," Journal of Fluid Mechanics, **14**, 481–495 (1962).

   D. Cylinder.

      Hermann Schlichting, *Boundary-Layer Theory,* 7th Edition, translated by J. Kestin. New York: McGraw-Hill Book Company, 1979, pp. 163–173.

   E. Wedges.

      Hermann Schlichting, *Boundary-Layer Theory,* 7th Edition, translated by J. Kestin. New York: McGraw-Hill Book Company, 1979, pp. 127–141.

F. Blasius series.

Hermann Schlichting, *Boundary-Layer Theory*, 7th Edition, translated by J. Kestin. New York: McGraw-Hill Book Company, 1979, pp. 168–173.

8. Graetz–Nusselt–Lévêque problem.

   Max Jakob, *Heat Transfer*, Volume I. New York: John Wiley & Sons, Inc., 1949. pp. 451–464.

9. Natural convection.

   George W. Morgan and W. H. Warner. "On Heat Transfer in Laminar Boundary Layers at High Prandtl Number." *Journal of the Aeronautical Sciences*, **23**, 937–948 (1956).

   BSL, 2nd Edition, pp. 316–319 and 339–353.

   Hermann Schlichting, *Boundary-Layer Theory*, 7th Edition, translated by J. Kestin. New York: McGraw-Hill Book Company, 1979, pp. 274–276.

   Veniamin G. Levich, *Physicochemical Hydrodynamics*. Englewood Cliffs, N. J.: Prentice-Hall, Inc., 1962, pp. 127–136.

   Andreas Acrivos, "A Theoretical Analysis of Laminar Natural Convection Heat Transfer to Non-Newtonian Fluids." *AIChE Journal*, **6**, 584–590 (1960).

10. High rates of mass transfer.

    BSL, 2nd Edition, pp. 703–716.

    Andreas Acrivos, "Mass Transfer in Laminar Boundary-Layer Flows with Finite Interfacial Velocities." *AIChE Journal*, **6**, 410–414 (1960).

    Andreas Acrivos, "The asymptotic form of the laminar boundary-layer mass-transfer rate for large interfacial velocities." *Journal of Fluid Mechanics*, **12**, 337–357 (1962).

    Donald R. Olander, "Rotating Disk Flow and Mass Transfer." *Journal of Heat Transfer*, **84C**, 185 (1962).

11. Lighthill transformation.

    M. J. Lighthill, "Contributions to the theory of heat transfer through a laminar boundary layer." *Proceedings of the Royal Society*, **A202**, 359–377 (1950).

    Andreas Acrivos. "Solution of the Laminar Boundary Layer Energy Equation at High Prandtl Numbers." *The Physics of Fluids*, **3**, 657–658 (1960).

12. Blasius series for mass transfer.
    Hermann Schlichting, *Boundary-Layer Theory*, 7th Edition, translated by J. Kestin. New York: McGraw-Hill Book Company, 1979, pp. 303–304.
    John Newman. "Blasius Series for Heat and Mass Transfer." *International Journal of Heat and Mass Transfer*, **9**, 705–709 (1966).

Notation

a	$= 0.51023$
$B(x)$	velocity derivative at the solid wall = $\sqrt{\nu}\partial v_x/\partial y$ at $y = 0$
c_i	concentration of diffusing species
c_0	concentration at solid surface
c_∞	concentration far from the surface
D	diffusion coefficient
$-D$	drag force
$\mathbf{e}_x$	unit vector in the x-direction
erf	error function
$\mathbf{g}$	gravitational acceleration
k	thermal conductivity
ℓ	characteristic length
$\mathbf{N}$	molar flux
Nu	$= \ell N / D(c_\infty - c_0)$, Nusselt number
p	thermodynamic pressure
P	dynamic pressure
Pe	$= v_\infty \ell / D$, the Péclet number
$\mathbf{r}$	position vector
$\mathcal{R}$	for an axisymmetric body is the distance of the surface from the axis of symmetry
r, ϕ, θ	spherical coordinates
r, θ, z	cylindrical coordinates
R	radius of sphere or pipe or a characteristic length
Re	$= v_\infty \ell / \nu$, Reynolds number
Sc	$= \nu / D$, Schmidt number
t	time
T	temperature
$U(x)$	velocity at outer edge of boundary layer

$\mathbf{v}$	mass-average velocity
$\mathbf{v}_\infty$	velocity far from body
v_r'	stretched velocity in boundary layer
x, y, z	rectangular coordinates
x	boundary-layer coordinate, distance measured along surface from front
y	boundary-layer coordinate, perpendicular distance from the surface
y'	stretched boundary-layer coordinate
Y	stretched, normal coordinate for diffusion layer at high Schmidt numbers
γ	Euler's constant
$\Gamma(4/3) = 0.89298$	
δ	boundary-layer thickness
ϵ	perturbation parameter
ζ	dimensionless variable (see Eq. 17.8)
η	dimensionless variable, variously used (see Eqs. 15.3, 20.9, 22.17, and 24.3)
Θ	$= (c_i - c_0)/(c_\infty - c_0)$, dimensionless concentration
μ	viscosity
ν	kinematic viscosity
ρ	density
τ	viscous stress
ψ	stream function
$\mathbf{\Omega}$	vorticity
Ω	rotation speed of disk
*	denotes dimensionless variable

SECTION C:
TRANSPORT IN TURBULENT FLOW

Chapter 33

Turbulent Flow and Hydrodynamic Stability

There are several aspects of turbulent flow, which might be considered in a fundamental course on fluid mechanics and mass transfer.

33.1 Time Averages of Equations of Motion, Continuity, and Convective Diffusion

Turbulent flow is characterized by rapid and random fluctuations of velocity, pressure, and concentration about their average values. One usually is interested in these fluctuations only in a statistical sense. Consequently, the first step in the study of turbulent flow usually involves an average of the equations presumed to describe the flow. This yields differential equations for certain average quantities, but with the involvement of higher-order averages. This procedure thus does not lead to any straightforward means of calculating any of the average quantities. The problem has a strong analogue in the kinetic theory of gases, where one is not interested in the details of the random motion of the molecules, but only in certain average, measurable quantities.

The Newman Lectures on Transport Phenomena
John Newman and Vincent Battaglia

ISBN 978-981-4774-27-7 (Hardcover), 978-1-315-10829-2 (eBook)
www.jennystanford.com

33.2 Hydrodynamic Stability

There are many situations for which a simple, laminar solution to the equation of motion can be found, but the actual flow is observed to be turbulent. This has led people to investigate the stability of the laminar flow; if the flow is disturbed by an infinitesimal amount, will the disturbance grow in time or distance or will the disturbance die away and leave the laminar flow? This analysis usually proceeds by linearizing the problem about the basic laminar solution. Sometimes the results agree with experimentally observed conditions of transition to turbulence, but sometimes there is a considerable discrepancy.

33.3 Eddy Viscosity, Eddy Diffusivity, and Universal Velocity Profile

Many practical problems of turbulence involve the region near a solid wall, since this is, in a sense, the origin of the turbulence and because it is in this region that we want to calculate shear stresses and rates of mass transfer. Experimental data have been studied extensively in order to draw some generalization about the behavior near the wall of the turbulent transport terms, these being the higher-order averages resulting from the averaging of the equations of motion and convective diffusion. This generalization takes the form of a universal law of velocity distribution near the wall, and the results can also be expressed in terms of the eddy viscosity and the eddy diffusivity, coefficients relating the turbulent transport terms to gradients of velocity and concentration. These coefficients are strong functions of the distance from the wall and, thus, are not fundamental fluid properties.

This type of information is frequently deduced from studies of fully developed pipe flow or certain simple boundary layers.

33.4 Application of These Results to Boundary Layers

These results are frequently used as the basis of a semi-empirical theory of turbulent flow, which can be applied to a wider class of problems than those that have been studied experimentally. Typical applications of this theory include the hydrodynamics of turbulent boundary layers, mass transfer in turbulent boundary layers, and the beginning of a mass-transfer section in fully developed pipe flow.

33.5 Statistical Theories of Turbulence

A fundamental treatment of turbulent flow has not yet been fully attained. The fundamental approaches tend to emphasize the decay of homogeneous turbulence, for example, created by a grid of bars being moved through the fluid. This is thus not a region near a solid wall. There has been some success predicting relationships among certain average quantities and confirming these predictions experimentally, but the fundamental problem still remains that formulas for statistical averages always involve higher-order averages.

Problems

33.1 For the velocity profile of Problems 15.2, 15.3, and 19.4, reflect on how to analyze the stability of the flow to small disturbances by answering the following questions:

(a) In the equation of motion, would one need to retain the derivative with respect to time?

(b) Suppose we represent the basic laminar flow by a superscript zero, as $\mathbf{v}^\circ$ and p°, and the disturbance by a prime, as $\mathbf{v}'$ and p'. Write out in vector form the equations governing the velocity and pressure, using this notation.

(c) Recall that the basic solution is an exact solution to the steady, axisymmetric fluid mechanical problem. Make the appropriate simplifications in the equations from Part (b) and write down the simplified equations.

(d) Can the stability of the basic flow to *small* disturbances be investigated accurately if we neglect in the equations of Part (c) all terms that are quadratic or of higher degree in the disturbance quantities (denoted by primes)?

(e) Is the problem for the disturbance then linear or nonlinear?

(f) Can the pressure disturbance p' be eliminated from the problem by taking the curl of the equation of motion?

(g) In steady axisymmetric problems, we sometimes introduce the Stokes stream function ψ such that

$$v_r = \frac{1}{r^2 \sin\theta}\frac{\partial \psi}{\partial \theta}, \quad v_\theta = \frac{-1}{r\sin\theta}\frac{\partial \psi}{\partial r}.$$

This has the advantage of replacing the vector unknown **v** by a scalar unknown ψ. If we do this in this problem, using ψ' for the disturbance, will we be able then to investigate the stability of the basic flow to arbitrary, small disturbances?

Chapter 34

Time Averages and Turbulent Transport

In practice, we frequently encounter turbulent flow. This is characterized by rapid and apparently random fluctuations of velocity and pressure about their mean values. These mean values can be defined by a time average, for example,

$$\bar{v}_z = \frac{1}{t_o}\int_t^{t+t_o} v_z dt. \tag{34.1}$$

The time t_o over which the average is taken should be long compared to the period of the fluctuations, which might be estimated as 0.01 s.

In laminar flow, the shear stress or momentum flux is given by Newton's law of viscosity, Eq. 2.3. However, in turbulent flow, there is an additional mechanism of momentum transfer. The random fluctuations of velocity tend to carry momentum toward regions of lower momentum. Thus, the total mean shear stress or momentum flux is the sum of a viscous shear stress and a turbulent momentum flux:

$$\bar{\tau}_{rz} = \bar{\tau}_{rz}^{(\ell)} + \bar{\tau}_{rz}^{(t)}, \tag{34.2}$$

where the viscous momentum flux $\bar{\tau}_{rz}^{(\ell)}$ is given by the time average of Eq. 2.3 and the turbulent momentum flux $\bar{\tau}_{rz}^{(t)}$ will be derived later in this chapter.

The Newman Lectures on Transport Phenomena
John Newman and Vincent Battaglia

ISBN 978-981-4774-27-7 (Hardcover), 978-1-315-10829-2 (eBook)
www.jennystanford.com

Far from a solid wall, momentum transfer by the turbulent mechanism predominates. However, near a solid wall, the turbulent fluctuations are damped, and viscous momentum transfer predominates, so that the shear stress at the wall is still given by

$$\tau_0 = -\mu \frac{\partial \bar{v}_z}{\partial r}\bigg|_{r=R} \tag{34.3}$$

for flow in a pipe of radius R. It seems reasonable that the turbulent fluctuations should be damped near the wall since the fluid cannot penetrate the wall.

The origin of the turbulent momentum flux is revealed by taking the time average of the equation of motion 2.2

$$\frac{\partial \rho \mathbf{v}}{\partial t} = -\nabla \cdot (\rho \mathbf{v}\, \mathbf{v}) - \nabla p - \nabla \cdot \boldsymbol{\tau}^{(\ell)} + \rho \mathbf{g}. \tag{34.4}$$

Here $\boldsymbol{\tau}^{(\ell)}$ denotes the same stress tensor that had previously been called $\boldsymbol{\tau}$ and is given by Eq. 2.3 for Newtonian fluids.

The deviation of a flow quantity from its time average is defined as follows for the velocity and the pressure:

$$\left.\begin{aligned} \mathbf{v} &= \overline{\mathbf{v}} + \mathbf{v}'. \\ p &= \overline{p} + \overline{p}'. \end{aligned}\right\} \tag{34.5}$$

We could call $\mathbf{v}'$ the velocity fluctuation or the fluctuating part of the velocity. Several rules of time averaging follow simply from the definition 34.1. The time average of a sum is equal to the sum of the time averages:

$$\overline{A+B} = \overline{A} + \overline{B}.$$

The time average of a derivative is equal to the derivative of the time average:

$$\overline{\frac{dA}{dx}} = \frac{d\overline{A}}{dx}.$$

In general, the time average of a nonlinear term will give more than one term. For example,

$$\overline{AB} = \overline{A}\,\overline{B} + \overline{A'B'}.$$

Of course, the time average of a fluctuation is zero

$$\overline{A'} = 0.$$

In this discussion, the fluid is assumed to have constant properties, ρ, μ, etc., since even with this assumption the turbulent-flow problem remains intractable and since incompressible fluids do exhibit turbulent flow. In fact, a compressible, laminar boundary layer may be more stable than an incompressible one. With this assumption, the time average of the equation of motion 34.4 yields

$$\frac{\partial \rho \bar{\mathbf{v}}}{\partial t} = -\nabla \cdot (\rho \bar{\mathbf{v}}\, \mathbf{v}) - \nabla \bar{p} - \nabla \cdot \left(\bar{\boldsymbol{\tau}}^{(\ell)} + \rho \overline{\mathbf{v}'\mathbf{v}'}\right) + \rho \mathbf{g}. \qquad (34.6)$$

The time-averaged continuity equation is

$$\nabla \cdot \bar{\mathbf{v}} = 0. \qquad (34.7)$$

The mean viscous stress is given by the time average of Eq. 2.3:

$$\bar{\boldsymbol{\tau}}^{(\ell)} = -\mu\left[\nabla \bar{\mathbf{v}} + (\nabla \bar{\mathbf{v}})^*\right]. \qquad (34.8)$$

These equations are the same as the equations before averaging except for the appearance of the term $-\nabla \cdot \left(\rho \overline{\mathbf{v}'\mathbf{v}'}\right)$ in the equation of motion 34.6. If we identify the turbulent momentum flux as

$$\bar{\mathbf{v}}^{(t)} = \rho \overline{\mathbf{v}'\mathbf{v}'} \qquad (34.9)$$

and write the total mean stress as

$$\bar{\boldsymbol{\tau}} = \bar{\boldsymbol{\tau}}^{(\ell)} + \bar{\boldsymbol{\tau}}^{(t)}, \qquad (34.10)$$

then the equation of motion becomes

$$\frac{\partial \rho \bar{\mathbf{v}}}{\partial t} = -\nabla \cdot (\rho \bar{\mathbf{v}} \bar{\mathbf{v}}) - \nabla \bar{p} - \nabla \cdot \bar{\boldsymbol{\tau}} + \rho \mathbf{g} \qquad (34.11)$$

and bears a strong resemblance to the equation before averaging.

These maneuvers illustrate the origin of the turbulent momentum flux or so-called Reynolds stress, given by Eq. 34.9. The turbulent mechanism of momentum transfer is somewhat similar to the molecular mechanism; one is due to random motion of molecules, and the other is due to random motion of larger, coherent aggregations of molecules. Recall the formula 3.6 obtained for the momentum transfer $\boldsymbol{\tau}^{(\ell)} + p\mathbf{I}$ in a pure fluid, in terms of molecular motion. In that formula, $\boldsymbol{\tau}^{(\ell)} + p\mathbf{I}$ represents the momentum flux due, partly, to deviations of individual molecular velocities from the average velocity $\mathbf{v}$. In that case, the average is over the velocity distribution function of the molecules. In Eq. 34.9, the Reynolds stress $\overline{\boldsymbol{\tau}}^{(t)}$ represents the mean momentum flux due to deviations of aggregations of molecules from the time-average velocity $\overline{\mathbf{v}}$.

The motivation for using an average is also similar in the two cases. Molecular motion and turbulent motion are both random phenomena, and we are interested only in certain statistical quantities of the flow but not in the detailed motion of each molecule or fluid aggregation. In both cases, the averaging does not introduce any significant approximations, but at the same time, it does not yield any quantitative information about the magnitude of the momentum flux, either $\overline{\boldsymbol{\tau}}^{(\ell)}$ or $\overline{\boldsymbol{\tau}}^{(t)}$. In the kinetic theory of gases, the process of molecular collisions is reexamined, and it is possible to derive Newton's law of viscosity 2.3 from the molecular average 3.6 and to express the viscosity in terms of molecular properties. Even so, much labor is involved, and the treatment is restricted to dilute gases of monatomic molecules.

In the same way, there is hope of developing a satisfactory statistical theory of turbulence, which will enable average quantities, such as the Reynolds stress, to be predicted. The difficulties encountered are somewhat similar to those of the kinetic theory of gases, but no reliable route to the solution has been found. In the absence of a fundamental theory, many people have written empirical expressions for $\overline{\boldsymbol{\tau}}^{(t)}$ with varying degrees of success. It should, perhaps, be emphasized that there is no simple relationship between turbulent stress and velocity derivatives, as there is for the viscous stress in a Newtonian fluid, where μ is a state function depending only on temperature, pressure, and composition.

For the consideration of heat and mass transfer in turbulent flow, one can average the energy equation or the material-balance equation. All nonlinear terms give rise to new terms in the averaged equation. The energy equation yields

$$\rho\hat{C}_V\left(\frac{\partial\overline{T}}{\partial t}+\overline{\mathbf{v}}\cdot\nabla\overline{T}\right)=-\nabla\cdot\left(\overline{\mathbf{q}}^{(\ell)}+\overline{\mathbf{q}}^{(t)}\right)-\overline{\boldsymbol{\tau}}^{(\ell)}:\nabla\overline{\mathbf{v}}-\overline{\boldsymbol{\tau}'^{(\ell)}:\nabla\mathbf{v}'}, \tag{34.12}$$

where

$$\overline{\mathbf{q}}^{(t)}=\rho\hat{C}_V\overline{\mathbf{v}'T'} \tag{34.13}$$

is the turbulent heat flux. Similarly, the material balance yields

$$\frac{\partial \bar{c}_i}{\partial t} + \bar{\mathbf{v}} \cdot \nabla \bar{c}_i = D\nabla^2 \bar{c}_i - \nabla \cdot \bar{\mathbf{J}}_i^{(t)} + \bar{R}_i, \tag{34.14}$$

where

$$\bar{\mathbf{J}}_i^{(t)} = \overline{\mathbf{v}' c_i'} \tag{34.15}$$

represents the mean turbulent mass flux due to the fluctuations of the concentration and the velocity from their mean values.

Chapter 35

Universal Velocity Profile and Eddy Viscosity

Our knowledge of turbulent flow is incomplete, and efforts to solve problems of turbulent flow by analytic means have not been completely successful. Experimental observations show that some important generalities can be made for turbulent flow near a wall. In particular, a "universal velocity profile" results if the mean tangential velocity is plotted against the distance from the wall, as shown in Fig. 35.1. This describes fully developed turbulent flow near a smooth wall and applies both to pipe flow and to turbulent boundary layers. The information is correlated by means of the shear stress τ_o at the wall:

$$v^+ = \frac{\overline{v}_x}{v_*}, \quad y^+ = \frac{y v_*}{\nu}, \quad v_* = \sqrt{\frac{\tau_o}{\rho}}. \tag{35.1}$$

Note that away from the wall, the mean velocity depends linearly on the logarithm of the distance, while near the wall, it increases linearly with the distance. The essential features of the curve are represented by the rough approximations

$$v^+ \approx y^+ \quad \text{for} \quad y^+ < 20 \tag{35.2}$$

and

$$v^+ \approx 2.5 \ln y^+ + 5.5 \quad \text{for} \quad y^+ > 20. \tag{35.3}$$

The Newman Lectures on Transport Phenomena
John Newman and Vincent Battaglia

ISBN 978-981-4774-27-7 (Hardcover), 978-1-315-10829-2 (eBook)
www.jennystanford.com

In the logarithmic region,

$$\bar{v}_x = 2.5\sqrt{\tau_o/\rho}\ \ln y + 2.5\sqrt{\tau_o/\rho}\ \ln \frac{\sqrt{\tau_o/\rho}}{\nu} + 5.5\sqrt{\tau_o/\rho}. \tag{35.4}$$

Here the term with the y-dependence of the velocity profile is independent of the viscosity; the viscosity of the fluid enters only into the additive constant.

From the data summarized in Fig. 35.1, it should be apparent that the Reynolds stress depends strongly on the distance from the wall. Many people have written empirical expressions for $\bar{\tau}_{xy}^{(t)}$ with varying degrees of success. Most find it convenient to introduce an "eddy viscosity" $\mu^{(t)}$ by the relation

$$\bar{\tau}_{xy}^{(t)} = -\mu^{(t)} \frac{\partial \bar{v}_x}{\partial y}. \tag{35.5}$$

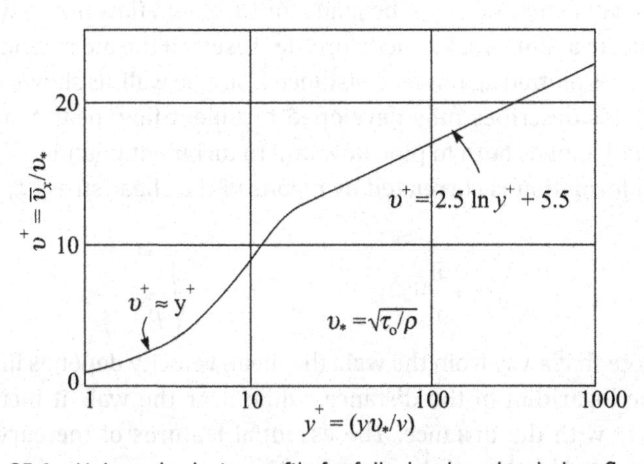

Figure 35.1 Universal velocity profile for fully developed turbulent flow.

The empirical results are then expressed in terms of the eddy viscosity. It should be noted that $\mu^{(t)}$ is not a fluid property and that the turbulent shear flow near a wall should not be expected to be isotropic. In other words, other components of the Reynolds stress probably require a different value of the eddy viscosity, even at the same distance from the wall.

The universal velocity profile of Fig. 35.1 probably applies only to a region near the wall where the shear stress is essentially constant, but not to the region near the center of a pipe, say, where the stress goes to zero. If we assume that the shear stress is constant over the region where the universal velocity profile is applicable, then we can obtain an idea of the variation of $\mu^{(t)}$ with distance from the wall.

$$\overline{\tau}_{xy} \approx -\tau_o = -\left(\mu + \mu^{(t)}\right)\frac{\partial \overline{v}_x}{\partial y} = -\frac{\mu + \mu^{(t)}}{\mu}\tau_o \frac{dv^+}{dy^+},$$

or

$$1 = \left[1 + \frac{\mu^{(t)}}{\mu}\right]\frac{dv^+}{dy^+}. \tag{35.6}$$

This shows that the ratio $\mu^{(t)}/\mu$ should also be a universal function of the wall variable y^+. Figure 35.2 is obtained by differentiation of the universal velocity profile of Fig. 35.1. It is not possible to obtain accurate values for $\mu^{(t)}$ near the wall by this method because in this region, $\mu^{(t)} \ll \mu$. However, this problem should not be of immediate concern since it is the sum of $\mu + \mu^{(t)}$, which enters into problems of fluid mechanics.

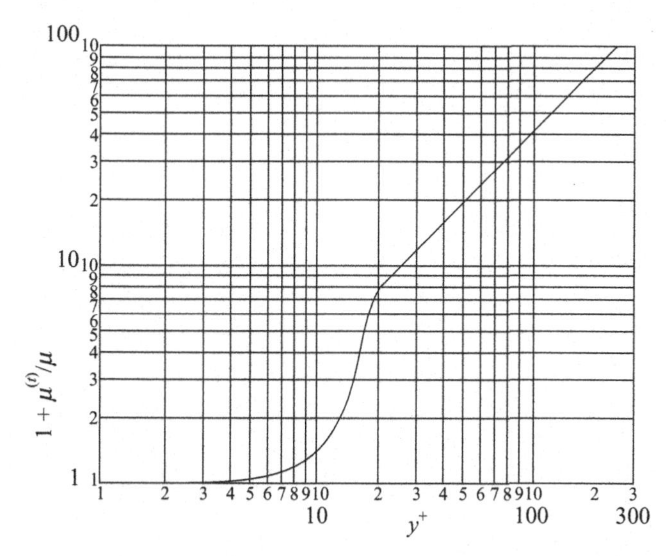

Figure 35.2 Representation of the eddy viscosity as a "universal" function of the distance from the wall.

The universal velocity profile is one of the few generalizations possible in turbulent flow, and it is widely applied in the analysis of problems for which experimental observations are not available. Before studying mass transfer in turbulent flow, we should illustrate in the next three chapters the application of the universal velocity profile to fluid-mechanical problems.

Chapter 36

Turbulent Flow in a Pipe

The concept of a universal velocity profile can be applied to fully developed turbulent flow in a pipe in two ways. First, one can assume that the universal velocity profile of Fig. 35.1 applies even though the total shear stress varies from τ_o at the wall to zero at the center. The other method is to assume that the total viscosity of Fig. 35.2 applies.

The average velocity in the pipe can be expressed as

$$\langle \bar{v}_z \rangle = \frac{2}{R^2} \int_0^R \bar{v}_z \, r \, dr. \tag{36.1}$$

In terms of the dimensionless velocity

$$v^+ = \bar{v}_z / \sqrt{\tau_o / \rho} \tag{36.2}$$

and the dimensionless distance from the wall

$$y^+ = y\sqrt{\tau_o / \rho} / \nu = (R - r)\sqrt{\tau_o / \rho} / \nu, \tag{36.3}$$

equation 36.1 can be rewritten as

$$\mathrm{Re} = 4 \int_0^{y^+_{\max}} v^+(y^+) \left(1 - \frac{y^+}{y^+_{\max}} \right) dy^+, \tag{36.4}$$

The Newman Lectures on Transport Phenomena
John Newman and Vincent Battaglia

ISBN 978-981-4774-27-7 (Hardcover), 978-1-315-10829-2 (eBook)
www.jennystanford.com

where $\mathrm{Re} = 2R\langle\bar{v}_z\rangle/\nu$ is the Reynolds number and $y^+_{max} = R\sqrt{\tau_o/\rho}/\nu$. By taking various values of y^+_{max} and using the universal velocity profile, one can use Eq. 36.4 to calculate the Reynolds number. One can then calculate the friction factor $f = 2\tau_o/\rho\langle\bar{v}_z\rangle^2$ since

$$y^+_{max} = \mathrm{Re}\sqrt{f/8}. \tag{36.5}$$

The results are given in Table 36.1. The velocity profile is then given by Fig. 35.1, where the values of y^+_{max} can be obtained from Table 36.1. It is a necessary consequence of this method of calculation that the velocity profile shows a peak at the center of the pipe, and the velocity derivative does not go to zero there, as it should.

Table 36.1 Friction factors for a pipe if the universal velocity profile applies.

Re	3000	10^4	3×10^4	10^5	3×10^5	10^6	3×10^6	10^7
$10^3 f$	12.01	7.96	5.91	4.48	3.58	2.88	2.40	1.99
y^+_{max}	116	315	815	2366	6349	1.90×10^4	5.19×10^4	1.58×10^5

It is interesting to use these results to calculate the total viscosity in the pipe by using the actual shear stress, which varies linearly from τ_o at the wall to zero at the center.

$$\frac{\bar{\tau}_{rz}}{\tau_o} = \frac{r}{R} = 1 - \frac{y^+}{y^+_{max}} = -\frac{\mu + \mu^{(t)}}{\tau_o}\frac{d\bar{v}_z}{dr} = \left(1 + \frac{\mu^{(t)}}{\mu}\right)\frac{dv^+}{dy^+}. \tag{36.6}$$

The results are shown in Fig. 36.1. It is a necessary consequence of this method of calculation that the total viscosity should go to zero at the center of the pipe, even though this is unreasonable. This means that the eddy viscosity is negative, $\mu^{(t)} = -\mu$, at the center.

The other method of calculation utilizes the actual shear-stress variation and the assumption that the total viscosity given by Fig. 35.2 applies. Equation 36.6 can be integrated to give the velocity profile:

$$v^+ = \int_0^{y^+} \frac{1 - y^+/y^+_{max}}{1 + \mu^{(t)}/\mu} dy^+. \tag{36.7}$$

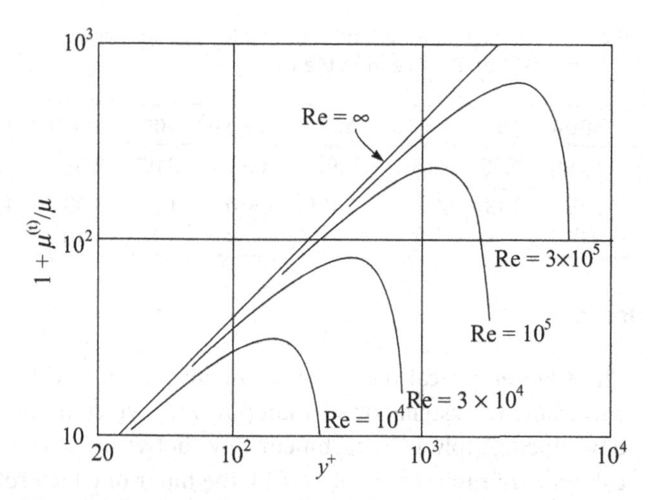

Figure 36.1 Total viscosity for pipe flow if the universal velocity profile applies all the way to the center of the pipe.

Again various values of y^+_{max} can be selected, the velocity profile can be calculated from Eq. 36.7, the Reynolds number from Eq. 36.4, and the friction factor from Eq. 36.5. These results are given in Table 36.2, and the velocity profiles are shown in Fig. 36.2. The gross results are not much different from those obtained by the first method, but the objectionable behavior at the center of the pipe is eliminated. This does not mean that the value of the total viscosity is correct at the center of the pipe, but it must be better than a value of zero.

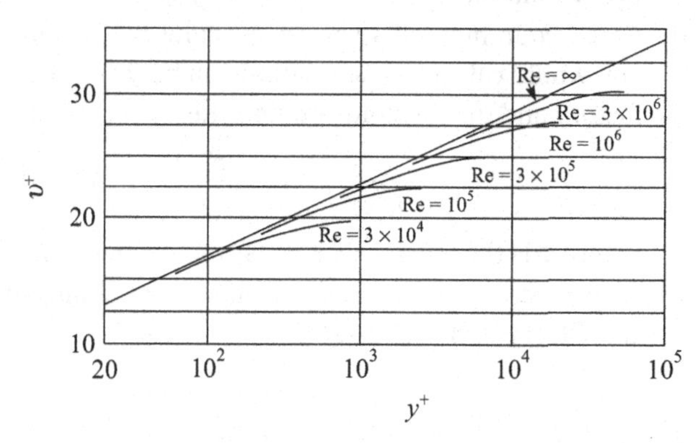

Figure 36.2 Velocity profiles for pipe flow for several values of the Reynolds number, calculated with the assumption that $\mu^{(t)}/\mu$ is a universal function of y^+.

Table 36.2 Friction factors for a pipe if the eddy viscosity follows a universal function of distance from the wall.

Re	3000	10^4	3×10^4	10^5	3×10^5	10^6	3×10^6	10^7
$10^3 f$	13.89	8.85	6.44	4.82	3.83	3.05	2.53	2.10
y^+_{max}	125	333	851	2455	6561	1.95×10^4	5.33×10^4	1.62×10^5

Problems

36.1 The semi-empirical theory of turbulent flow purports to be applicable to systems other than pipes for which it has been developed. Apply it to turbulent flow between two circular cylinders of radii R and κR ($\kappa < 1$), the inner of which rotates with an angular velocity Ω and the outer of which is stationary.

(a) Determine the distribution of mean stress in the region between the circular cylinders.

(b) Outline in words how you would apply the concepts of the universal velocity profile or the universal law of eddy viscosity to determine the mean velocity in this region. Make sure that your statement is complete and indicates clearly what parameters are taken to be independent (i.e., specified) and which are to be calculated.

(c) Discuss the relative merits of using the universal velocity profile and the universal law of eddy viscosity.

(d) Carry your method through to show by formulas how you would calculate the relationship between the friction factor f and the Reynolds number Re:

$$f = \frac{2\tau_i}{\rho \kappa^2 R^2 \Omega^2}, \quad \text{Re} = \frac{R^2 \Omega}{\nu} \kappa (1 - \kappa),$$

where τ_i is the mean shear stress on the inner cylinder.

(e) Discuss any reservations you may have about applying this theory to this problem.

Chapter 37

Integral Momentum Method for Boundary Layers

For a steady, turbulent, two-dimensional boundary layer, the equation of motion can be approximated by

$$\rho\left(\bar{v}_x\frac{\partial \bar{v}_x}{\partial x}+\bar{v}_y\frac{\partial \bar{v}_x}{\partial y}\right)=\rho U\frac{dU}{dx}+\frac{\partial}{\partial y}(\mu+\mu^{(t)})\frac{\partial \bar{v}_x}{\partial y}, \tag{37.1}$$

and the boundary conditions are

$$\bar{v}_x=0 \text{ at } y=0 \text{ and } \bar{v}_x=U(x) \text{ at } y=\infty. \tag{37.2}$$

By use of the continuity equation

$$\frac{\partial \bar{v}_x}{\partial x}+\frac{\partial \bar{v}_y}{\partial y}=0, \tag{37.3}$$

the left side of Eq. 37.1 can be rewritten as

$$\frac{\partial \bar{v}_x\bar{v}_x}{\partial x}+\frac{\partial \bar{v}_y\bar{v}_x}{\partial y}-U\frac{dU}{dx}=\frac{\partial}{\partial y}\frac{\mu+\mu^{(t)}}{\rho}\frac{\partial \bar{v}_x}{\partial y}. \tag{37.4}$$

Integration with respect to y gives

$$\left.\frac{\mu+\mu^{(t)}}{\rho}\frac{\partial \bar{v}_x}{\partial y}\right|_0^y=\frac{d}{dx}\int_0^y \bar{v}_x\bar{v}_x\,dy+\bar{v}_y\bar{v}_x-\frac{dU}{dx}\int_0^y U\,dy. \tag{37.5}$$

The Newman Lectures on Transport Phenomena
John Newman and Vincent Battaglia

ISBN 978-981-4774-27-7 (Hardcover), 978-1-315-10829-2 (eBook)
www.jennystanford.com

We can write

$$\frac{d}{dx}\int_0^y \bar{v}_x\bar{v}_x dy = \frac{d}{dx}\int_0^y \bar{v}_x(\bar{v}_x - U)\,dy + \int_0^y\left(\bar{v}_x\frac{dU}{dx} + U\frac{\partial \bar{v}_x}{\partial x}\right)dy$$

$$= \frac{d}{dx}\int_0^y \bar{v}_x(\bar{v}_x - U)dy + \frac{dU}{dx}\int_0^y \bar{v}_x dy - U\int_0^y \frac{\partial \bar{v}_y}{\partial y}dy$$

$$= \frac{d}{dx}\int_0^y \bar{v}_x(\bar{v}_x - U)dy + \frac{dU}{dx}\int_0^y \bar{v}_x dy - U\bar{v}_y.$$

Hence, Eq. 37.5 becomes

$$\frac{\mu + \mu^{(t)}}{\rho}\frac{\partial \bar{v}_x}{\partial y}\bigg|_0^y = \frac{d}{dx}\int_0^y \bar{v}_x(\bar{v}_x - U)\,dy + \frac{dU}{dx}\int_0^y (\bar{v}_x - U)dy + \bar{v}_y(\bar{v}_x - U)\bigg|_0^y. \tag{37.6}$$

If now we let the upper limit of integration extend to infinity, we obtain the "integral momentum equation:"

$$\frac{\tau_0}{\rho} = \frac{d}{dx}\int_0^\infty \bar{v}_x(U - \bar{v}_x)\,dy + \frac{dU}{dx}\int_0^\infty (U - \bar{v}_x)\,dy, \tag{37.7}$$

where τ_0 is the shear stress at the wall. This equation expresses the fact that the momentum deficiency in the boundary layer (relative to the external flow) is due to the shear stress at the wall.

Let us assume that the universal velocity profile applies over a region of thickness $\delta(x)$, at which value of y the velocity attains the value $U(x)$. In order to use this idea, it is necessary to determine the variation of shear stress $\tau_0(x)$ along the wall, since this enters the correlation for the universal velocity profile. With this assumption, Eq. 37.7 takes the form

$$v_*^2 = \delta U\frac{dU}{dx} - \nu\frac{dv_*}{dx}\int_0^{\delta+} v^{+2}dy^+. \tag{37.8}$$

This provides an ordinary differential equation for $v_*(x)$, where δ and δ^+ are determined by the relations

$$U(x) = v_*(x)v^+(\delta^+) \text{ and } \delta^+ = \delta v_*/\nu. \tag{37.9}$$

The method outlined here allows an approximate treatment of turbulent boundary layers based on the universal velocity profile. The boundary-layer equation 37.1 is not satisfied in detail, but only in the integral form 37.7, with an assumed form for the velocity profile. The method is somewhat similar to the first one used for pipe flow in Chapter 36, where the universal velocity profile was assumed to apply all the way to the center of the pipe.

Problems

37.1 For turbulent flow past a flat plate parallel to the free stream, discuss how to obtain approximate expressions for the shear stress distributions on the basis of the integral momentum method. Assume that the boundary layer is turbulent from the beginning of the plate. Obtain an expression for the stress distribution close to the leading edge and compare this numerically with the result of Blasius for laminar flow. Seek an expression for τ_o far downstream in the fully turbulent region by using the universal velocity profile in the approximate form

$$v^+ = 5.3\,(y^+)^{1/7}.$$

37.2 The turbulent boundary-layer equations are given at the beginning of this chapter—the first three equations. Discuss the justification of these equations on the basis of order-of-magnitude arguments. You may restrict yourself to a finite flat plate of length L. The velocity $U(x)$ outside the boundary layer is not necessarily constant, but it can be taken to be of the order of v_∞, a characteristic velocity. The Reynolds number $v_\infty L/\nu$ is a large parameter. For order-of-magnitude purposes, the Reynolds stress can be considered to be related to velocity derivatives by a formula similar to that for the viscous stress but with the viscosity replaced by the eddy viscosity. The latter can be taken to be given by the universal law of the wall. The shear stress at the wall can be expressed by the result of Problem 37.1:

$$\tau_o = 0.058\,\rho v_\infty^2 \left(\frac{v_\infty x}{\nu}\right)^{-0.2}.$$

In your analysis, consider the following:

(a) The order of magnitude of the pressure difference across the thickness of the boundary layer.

(b) The order of magnitude of neglected components of the Reynolds stress and their derivatives that might appear in the boundary-layer equation of motion or the assessment of the pressure difference in Part (a).

(c) Other boundary conditions that might be appropriate or required.

Chapter 38

Use of Universal Eddy Viscosity for Turbulent Boundary Layers

The boundary-layer equation 37.1 provides a satisfactory basis for treating turbulent boundary layers if some additional information about the eddy viscosity $\mu^{(t)}$ can be supplied. In a variation of the semi-empirical theory, one can assume that $\mu^{(t)}$ is given as a universal function of the distance from the wall, as in Fig. 35.2. This, again, requires a knowledge of the shear stress $\tau_0(x)$ at the wall, but this can be determined in the course of solving the problem. Thus, from the parabolic nature of the differential equations, one can start at the beginning of the boundary layer (or at the beginning of the turbulent part of the boundary layer) and integrate the equations downstream, numerically if necessary. At each station or value of x, one evaluates the shear stress at the wall and uses this value to determine the eddy viscosity for all values of y.

This method is somewhat similar to the second method used for pipe flow in Chapter 36, where the eddy viscosity given by Fig. 35.2 was assumed to apply all the way to the center of the pipe. We should expect that we can eliminate the discontinuity in the velocity derivative at $y = \delta$ that results in the method of Chapter 37.

The Newman Lectures on Transport Phenomena
John Newman and Vincent Battaglia

ISBN 978-981-4774-27-7 (Hardcover), 978-1-315-10829-2 (eBook)
www.jennystanford.com

Chapter 39

Mass Transfer in Turbulent Flow

Next one might address oneself to the problem of estimating mass-transfer rates in turbulent flow, but with the use of as little additional information as possible. One usually starts with the assumption that the eddy diffusivity $D^{(t)}$, defined by the relation

$$\bar{J}_{iy}^{(t)} = \overline{c_i' v_y'} = -D^{(t)} \frac{\partial \bar{c}_i}{\partial y}, \tag{39.1}$$

is related to or equal to the eddy kinematic viscosity

$$D^{(t)} = \nu^{(t)} = \mu^{(t)}/\rho. \tag{39.2}$$

This is based on the idea that transfer of momentum and mass is similar, whether it is by a molecular or by a turbulent mechanism. Thus Fig. 35.2 can be used to get information about the variation of the eddy diffusivity in fully developed, turbulent flow near a wall. As stated earlier, $\nu^{(t)}$ is much smaller than ν very close to the wall, and Fig. 35.2 gives no information about it. But for mass transfer at large Schmidt numbers, it is necessary to know $D^{(t)}$ closer to the wall. Thus, even if $\nu^{(t)} = D^{(t)}$, we can have at some distance $D^{(t)} >> D$ even where $\nu^{(t)} << \nu$ if Sc is large.

Thus, much of our information about $D^{(t)}$ and possibly $\nu^{(t)}$ near the wall comes from mass-transfer experiments. Actually it probably results more from fitting average mass-transfer rates at the wall

The Newman Lectures on Transport Phenomena
John Newman and Vincent Battaglia

ISBN 978-981-4774-27-7 (Hardcover), 978-1-315-10829-2 (eBook)
www.jennystanford.com

than from examination in detail of actual concentration profiles. If, for mass transfer in a pipe, we measure the rate by means of the Stanton number

$$\mathrm{St} = \frac{D}{\langle \bar{v}_z \rangle \Delta c_i} \frac{\partial \bar{c}_i}{\partial r}\bigg|_{r=R} = \frac{\mathrm{Nu}}{\mathrm{ReSc}}, \tag{39.3}$$

then the dependence of the Stanton number on the Schmidt number at large Schmidt numbers is determined by the variation of the eddy diffusivity close to the wall as follows:

$D^{(t)}$ near the wall	**St for large Sc**
$D^{(t)} \propto y^2$	$\mathrm{St} \propto \mathrm{Sc}^{-1/2}$
y^3	$\mathrm{Sc}^{-2/3}$
y^4	$\mathrm{Sc}^{-3/4}$

It is easy to show that $\nu^{(t)}$ must go to zero as y^3 or a higher power of y but cannot vary as y^2. If v'_x and v'_y are expanded in power series near the wall, then v'_x is proportional to y and, by the continuity equation, v'_y is proportional to y^2. Hence, $\bar{\tau}^{(t)}_{xy} = \rho \overline{v'_x v'_y}$ is proportional to y^3, and the same must be true of $\nu^{(t)}$.

Statements about the similarity between momentum and mass or heat transfer in fully developed flow are called "analogies." The first of these described an analogy only between the average mass-transfer rate and the shear stress at the wall. Where f is the friction factor, Reynolds (1874) obtained

$$\mathrm{St} = \frac{f}{2}, \tag{39.4}$$

which is good only for Schmidt numbers close to one. Chilton and Colburn used the expression [1]

$$\mathrm{St} = \frac{1}{2} f\, \mathrm{Sc}^{-2/3}, \tag{39.5}$$

which is very good for a wide range of Schmidt numbers greater than 0.5. Here f is related to the viscous stress and should not include any pressure drag.

Other, more detailed "analogies" relate $D^{(t)}$ to $\nu^{(t)}$ and then relate $\nu^{(t)}$ to the distance from the wall in the form of $\nu^{(t)}/\nu$ as a function of y^+. Early theories said that $\nu^{(t)} = 0$ near the wall, but actually the turbulence level varies continuously and goes to zero only at the wall. Since then the "analogies" have proliferated. Levich (1942) had $D^{(t)}$ proportional to y^3 but later (1944) changed this to y^4 [2]. See also Ref. [3]. Lin, Moulton, and Putnam (1953) used y^3 [4]. Deissler's (1955) expression varies as y^4 near the wall [5]. These attempts at curve fitting have been reviewed by Sherwood (1959) [6]. The mass-transfer results at high Schmidt numbers do not seem to be sufficiently accurate or sensitive to resolve the question of whether $D^{(t)}$ varies as y^3 or y^4 near the wall. (See also Ref. [7])

The goal of all this work is to determine how the eddy diffusivity depends on distance from the wall, that is, to relate $D^{(t)}/\nu$ to y^+, the dimensionless distance from the wall. This is based on the universal velocity profile and on information gleaned from mass-transfer experiments. The result is sketched in Fig. 39.1. For this purpose, we have used the expression of Wasan, Tien, and Wike [8], who use

$$\left.\begin{aligned} v^+ &= y^+ - A_1 y^{+4} + A_2 y^{+5} && \text{for } y^+ \le 20, \\ v^+ &= 2.5 \ln y^+ + 5.5 && \text{for } y^+ \ge 20. \end{aligned}\right\} \tag{39.6}$$

The constants A_1 and A_2 are selected so that v^+ and its derivative are continuous at $y^+ = 20$:

$$A_1 = 1.0972 \times 10^{-4} \quad \text{and} \quad A_2 = 3.295 \times 10^{-6}. \tag{39.7}$$

The corresponding expressions for the eddy diffusivity are

$$\left.\begin{aligned} \frac{D^{(t)}}{\nu} &= \frac{4A_1 y^{+3} - 5A_2 y^{+4}}{1 - 4A_1 y^{+3} + 5A_2 y^{+4}} && \text{for } y^+ \le 20, \\ \frac{D^{(t)}}{\nu} &= \frac{y^+}{2.5} - 1 && \text{for } y^+ \ge 20. \end{aligned}\right\} \tag{39.8}$$

The concept of the universal velocity profile and the variation of eddy diffusivity with distance from the wall as given in Fig. 39.1 form the basis of a semi-empirical theory widely used to calculate mass-transfer rates in turbulent boundary layers, near the beginning of a mass-transfer section in a pipe, and for similar problems.

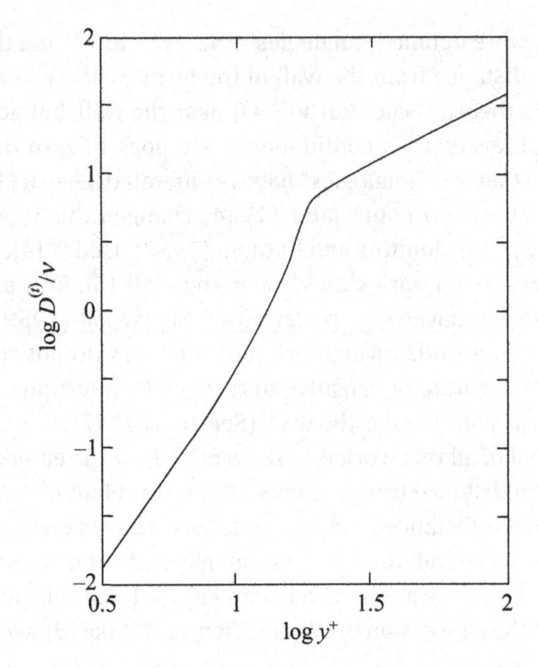

Figure 39.1 Variation of eddy diffusivity near a wall for fully developed turbulent flow.

Problem

39.1 For mass transfer in fully developed, turbulent pipe flow, we found that the dependence of the Stanton number on the Schmidt number at high Schmidt numbers is determined by the nature of the turbulent fluctuations very close to the wall, as follows:

Behavior of $D^{(t)}$ near the wall	Behavior of St at high Sc
$D^{(t)} \propto y^2$	$\mathrm{St} \propto \mathrm{Sc}^{-1/2}$
y^3	$\mathrm{Sc}^{-2/3}$
y^4	$\mathrm{Sc}^{-3/4}$

In early work on turbulent mass transfer, it was assumed that

$$D^{(t)} = 0 \text{ for } y^+ < 5.$$

For high Schmidt numbers, what result does this give for the Stanton number? Include the dependence on the friction factor or the Reynolds number and any numerical constants. Contrast this result with Reynolds analogy and the Chilton–Colburn relation.

References

1. T. H. Chilton and A. P. Colburn. "Mass Transfer (Absorption) Coefficients. Prediction from Data on Heat Transfer and Fluid Friction." *Industrial and Engineering Chemistry*, **26**, 1183–1187 (1934).
2. B. Levich. "The Theory of Concentration Polarization." *Acta Phyaicochimica URSS*, **17**, 257–307 (1942); **19**, 117–132 (1944).
3. Veniamin G. Levich. *Physicochemical Hydrodynamics*. Englewood Cliffs, N. J.: Prentice-Hall, Inc., 1962.
4. C. S. Lin, R. W. Moulton, and G. L. Putnam. "Mass Transfer between Solid Wall and Fluid Streams. Mechanism and Eddy Distribution Relationships in Turbulent Flow." *Industrial and Engineering Chemistry*, **45**, 636–640 (1953).
5. Robert G. Deissler. "Analysis of Turbulent Heat Transfer, Mass Transfer, and Friction in Smooth Tubes at High Prandtl and Schmidt Numbers." Report 1210. *Forty-first Annual Report of the National Advisory Committee for Aeronautics-1955*. Washington: United States Government Printing Office, 1957.
6. T. K. Sherwood. "Mass, Heat, and Momentum Transfer between Phases." *Chemical Engineering Progress Symposium Series*, **55** (25) pp. 71–85 (1959).
7. Davis Watson Hubbard. *Mass Transfer in Turbulent Flow at High Schmidt Numbers*. Dissertation, University of Wisconsin, 1964.
8. D. T. Wasan, C. L. Tien, and C. R. Wilke. "Theoretical Correlation of Velocity and Eddy Viscosity for Flow Close to a Pipe Wall." *AIChE Journal*, **9**, 567–568 (1963).

Chapter 40

Mass Transfer in Turbulent Pipe Flow

Let us first apply the semi-empirical theory of turbulent mass transfer to fully developed, turbulent flow in a pipe, since it is a comparison with the results of such experiments that provides the necessary additional information on the variation of the eddy diffusivity near the wall. It is convenient to consider fully developed, turbulent mass transfer with a constant flux at the pipe wall. The differential equation for mass transfer 34.14 takes the form

$$\bar{v}_z \frac{\partial \bar{c}_i}{\partial z} = -\nabla \cdot \bar{\mathbf{J}}_i = -\frac{1}{r}\frac{\partial}{\partial r}(r\bar{J}_{ir}) - \frac{\partial \bar{J}_{iz}}{\partial z}. \tag{40.1}$$

For fully developed mass transfer with a constant wall flux, $\partial \bar{c}_i / \partial z$ is a constant independent of r and z, and $\partial \bar{J}_{iz} / \partial z = 0$. Therefore, Eq. 40.1 can be integrated to yield

$$-r\bar{J}_{ir} = \frac{\partial \bar{c}_i}{\partial z}\int_0^r r\,\bar{v}_z\, dr. \tag{40.2}$$

The constant $\partial \bar{c}_i / \partial z$ is thus related to the flux J_o at the wall by

$$-RJ_o = \frac{\partial \bar{c}_i}{\partial z}\int_0^R r\bar{v}_z\, dr, \tag{40.3}$$

The Newman Lectures on Transport Phenomena
John Newman and Vincent Battaglia

ISBN 978-981-4774-27-7 (Hardcover), 978-1-315-10829-2 (eBook)
www.jennystanford.com

and Eq. 40.2 becomes

$$\frac{\bar{J}_{ir}}{J_o} = \frac{R}{r}\int_0^r r\,\bar{v}_z\,dr \Bigg/ \int_0^R r\,\bar{v}_z\,dr$$

$$= \frac{1}{1 - y^+/y^+_{max}} \int_{y^+}^{y^+_{max}} v^+\left(1 - \frac{y^+}{y^+_{max}}\right)dy^+ \Bigg/ \int_0^{y^+_{max}} v^+\left(1 - \frac{y^+}{y^+_{max}}\right)dy^+$$

or

$$\frac{\bar{J}_{ir}}{J_o} = \frac{4/\text{Re}}{1 - y^+/y^+_{max}} \int_{y^+}^{y^+_{max}} v^+\left(1 - \frac{y^+}{y^+_{max}}\right)dy^+, \qquad (40.4)$$

where Eq. 36.4 has been used.

The variation of $\bar{J}_{ir}/J_o$ is sketched in Fig. 40.1. For this purpose, we have used a velocity distribution given by

$$\frac{\bar{v}_z}{v_{z,max}} = \left(1 - \frac{r}{R}\right)^{1/7}, \qquad (40.5)$$

although this is not essential for illustrating the features of the curve. The dimensionless flux rises above unity and has a slope of +1 at the wall. It then decreases approximately linearly and goes to zero at the center of the pipe. Such a curve can be constructed for any Reynolds number from the velocity distribution calculated by means of Eq. 36.7 (the second method of Chapter 36).

From a knowledge of the distributions of flux and eddy diffusivity, the concentration distribution can now be determined from the equation

$$\bar{J}_{ir} = -\left(D + D^{(t)}\right)\frac{\partial \bar{c}_i}{\partial r}. \qquad (40.6)$$

In terms of the dimensionless concentration

$$\Theta^+ = \frac{v_*}{J_o}\left[\bar{c}_i - \bar{c}_i(y=0)\right], \qquad (40.7)$$

Eq. 40.6 takes the form

$$\frac{\bar{J}_{ir}}{J_o} = \frac{D + D^{(t)}}{\nu}\frac{d\Theta^+}{dy^+}, \qquad (40.8)$$

which can be integrated to give

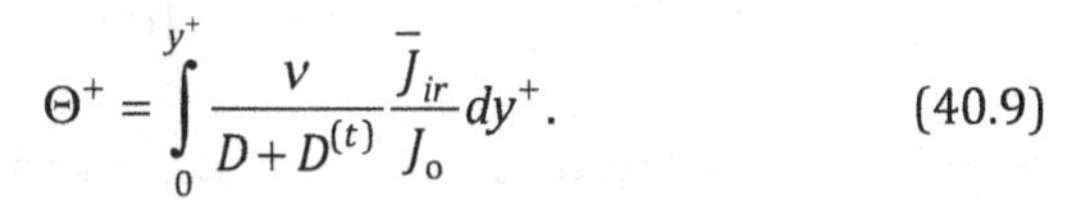

$$\Theta^+ = \int_0^{y^+} \frac{\nu}{D+D^{(t)}} \frac{\bar{J}_{ir}}{J_o} dy^+. \tag{40.9}$$

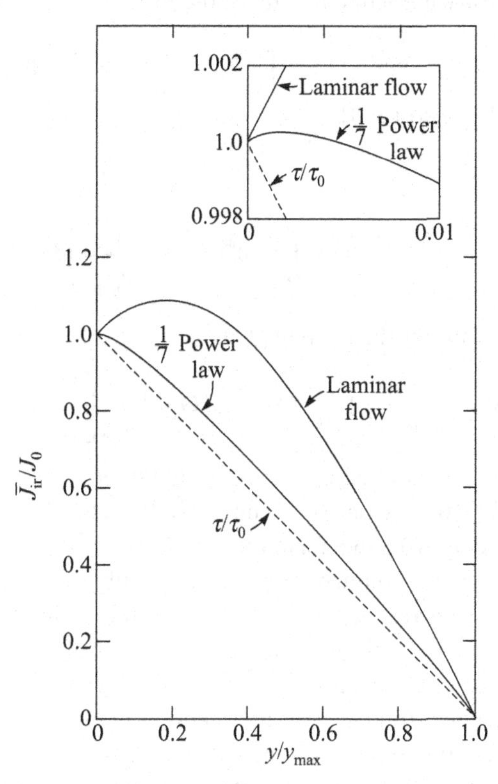

Figure 40.1 Variation of the radial flux divided by the wall flux with distance from the wall. The dashed line shows the stress divided by the stress at the wall.

It is now of interest to calculate the Nusselt number or the Stanton number for fully developed turbulent mass transfer in a pipe for comparison with experimental results. The Nusselt number is defined here as

$$\mathrm{Nu} = \frac{J_o}{D} \frac{2R}{c_{avg} - \bar{c}_i(y=0)}, \tag{40.10}$$

where c_{avg}, sometimes called the cup-mixing average concentration, is defined as

$$c_{\text{avg}} = \int_0^R \bar{v}_z \bar{c}_i \, r\, dr \bigg/ \int_0^R \bar{v}_z \, r\, dr. \tag{40.11}$$

Expressed in terms of the dimensionless variables appropriate to the semi-empirical theory, this takes the form

$$\begin{aligned} c_{\text{avg}} - \bar{c}_i(y=0) &= \frac{J_0}{v_*} \frac{\displaystyle\int_0^{y^+_{\max}} v^+ \Theta^+ \left(1 - \frac{y^+}{y^+_{\max}}\right) dy^+}{\displaystyle\int_0^{y^+_{\max}} v^+ \left(1 - \frac{y^+}{y^+_{\max}}\right) dy^+} \\ &= \frac{4J_0}{v_* \text{Re}} \int_0^{y^+_{\max}} v^+ \Theta^+ \left(1 - \frac{y^+}{y^+_{\max}}\right) dy^+. \end{aligned} \tag{40.12}$$

Substitution into Eq. 40.10 then gives

$$\text{St} = \frac{\text{Nu}}{\text{ReSc}} = y^+_{\max} \bigg/ 2\int_0^{y^+_{\max}} v^+ \Theta^+ \left(1 - \frac{y^+}{y^+_{\max}}\right) dy^+. \tag{40.13}$$

The Stanton number can thus be obtained by evaluating the integral in Eq. 40.13, where Θ^+ is obtained from Eq. 40.9. The results are shown in Fig. 40.2 in a form that facilitates comparison with Eq. 39.5 of Chilton and Colburn. The Stanton number can be recovered, if desired, from the friction factors in Table 36.2. Very good agreement with the Chilton–Colburn equation is obtained for Sc = 1, and the agreement becomes better at higher Reynolds numbers. The differences reflect the fact that mass and momentum transfers in a pipe are not completely analogous even at Sc = 1. The discrepancy becomes greater at higher Schmidt numbers, particularly at higher Reynolds numbers.

It is instructive to derive the asymptotic expression for the Stanton number in the limit of large Schmidt numbers. For large Schmidt numbers, the diffusion coefficient is very small, and the concentration variation occurs in a thin diffusion layer near the wall, much as in laminar flow problems. This has the first consequence that the concentration is uniform over most of the cross section of the pipe, and Eq. 40.13 can be approximated by

$$\text{St} = y^+_{\max} \bigg/ 2\Theta^+_{\max} \int_0^{y^+_{\max}} v^+ \left(1 - \frac{y^+}{y^+_{\max}}\right) dy^+ = \frac{2y^+_{\max}}{\text{Re}\Theta^+_{\max}} = \frac{\sqrt{f/2}}{\Theta^+_{\max}}. \tag{40.14}$$

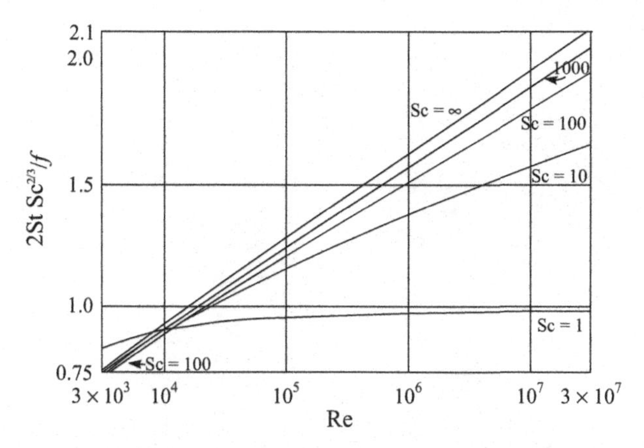

Figure 40.2 Calculated rates of fully developed mass transfer for turbulent pipe flow, compared with the expression of Chilton and Colburn.

Next, $\Theta^+_{\max}$ can be obtained from Eq. 40.9, with regard for the behavior of the eddy viscosity only in the region near the wall and with the flux approximated by the value at the wall:

$$\Theta^+_{\max} = \int_0^{y^+_{\max}} \frac{\nu}{D + D^{(t)}} dy^+ = \int_0^{y^+_{\max}} \frac{dy^+}{1/\mathrm{Sc} + 4A_1 y^{+3}}. \quad (40.15)$$

Here $D^{(t)}/\nu$ has been approximated by the appropriate form near the wall as given by Eq. 39.8. Equation 40.15 then takes the form

$$\Theta^+_{\max} = \frac{\mathrm{Sc}^{2/3}}{(4A_1)^{1/3}} \int_0^{\infty} \frac{dx}{1+x^3} = \frac{\mathrm{Sc}^{2/3}}{(4A_1)^{1/3}} \Gamma\left(\frac{4}{3}\right)\Gamma\left(\frac{2}{3}\right) = 1.2092 \frac{\mathrm{Sc}^{2/3}}{(4A_1)^{1/3}}. \quad (40.16)$$

Thus, the Stanton number approaches

$$\mathrm{St} \to \frac{(4A_1)^{1/3}}{1.2092} \sqrt{\frac{f}{2}} \mathrm{Sc}^{-2/3} \text{ as } \mathrm{Sc} \to \infty. \quad (40.17)$$

This has a dependence on the Reynolds number different from that of the Chilton–Colburn expression since it involves the square root of the friction factor. Agreement with the Chilton–Colburn expression is obtained when $(4A_1)^{1/3}/1.2092 = \sqrt{f/2}$, and this occurs at a Reynolds number of about 15,000.

Chapter 41

Mass Transfer in Turbulent Boundary Layers

For a steady, turbulent, two-dimensional boundary layer, the equation of motion is given by Eq. 37.1, and the equation of continuity by Eq. 37.3. The boundary-layer form of the time-averaged equation of convective diffusion is

$$\bar{v}_x \frac{\partial \bar{c}_i}{\partial x} + \bar{v}_y \frac{\partial \bar{c}_i}{\partial x} = D\frac{\partial^2 \bar{c}_i}{\partial y^2} - \frac{\partial \overline{v'_y c'_i}}{\partial y} = \frac{\partial}{\partial y}\left[\left(D + D^{(t)}\right)\frac{\partial \bar{c}_i}{\partial y}\right]. \quad (41.1)$$

If we know something about the fluid mechanics of the turbulent flow, what can we deduce about the mass-transfer problem?

Spalding showed us one way to solve the mass-transfer, boundary-layer equation for turbulent flow [1]. For a constant concentration c_0 at the wall, he suggested the variables

$$\Theta = \frac{\bar{c}_i - c_0}{c_\infty - c_0} \quad \text{and} \quad \psi = \int_0^y \bar{v}_x dy, \quad (41.2)$$

so that

$$\bar{v}_x = \frac{\partial \psi}{\partial y} \text{ and } \bar{v}_y = -\frac{\partial \psi}{\partial x}. \quad (41.3)$$

Let us transform the independent variables to x and ψ:

The Newman Lectures on Transport Phenomena
John Newman and Vincent Battaglia

ISBN 978-981-4774-27-7 (Hardcover), 978-1-315-10829-2 (eBook)
www.jennystanford.com

$$\left(\frac{\partial\Theta}{\partial x}\right)_y = \left(\frac{\partial\Theta}{\partial x}\right)_\psi \left(\frac{\partial x}{\partial x}\right)_y + \left(\frac{\partial\Theta}{\partial \psi}\right)_x \left(\frac{\partial \psi}{\partial x}\right)_y = \left(\frac{\partial\Theta}{\partial x}\right)_\psi - \bar{\upsilon}_y \left(\frac{\partial\Theta}{\partial \psi}\right)_x .$$

$$\left(\frac{\partial\Theta}{\partial x}\right)_x = \left(\frac{\partial\Theta}{\partial \psi}\right)_x \left(\frac{\partial \psi}{\partial y}\right)_x + \left(\frac{\partial\Theta}{\partial x}\right)_\psi \left(\frac{\partial x}{\partial y}\right)_x = \bar{\upsilon}_x \frac{\partial\Theta}{\partial \psi} .$$

The diffusion equation 41.1 becomes

$$\frac{\partial\Theta}{\partial x} = \frac{\partial}{\partial \psi}\left[(D + D^{(t)})\bar{\upsilon}_x \frac{\partial\Theta}{\partial \psi}\right]. \tag{41.4}$$

Now introduce the following independent variables, appropriate to the universal velocity profile,

$$\upsilon^+ = \frac{\bar{\upsilon}_x}{\upsilon_*} \text{ and } x^+ = \frac{1}{\nu}\int_{x_0}^{x} \upsilon_*(x)\,dx, \tag{41.5}$$

where $\upsilon_* = \sqrt{\tau_0(x)/\rho}$ and x_0 denotes the beginning of the mass-transfer section and may not coincide with the beginning of the turbulent, hydrodynamic boundary layer. If we let $y^+ = y\upsilon_*/\nu$, then we can relate υ^+ to ψ:

$$\psi = \int_0^y \bar{\upsilon}_x\, dy = \nu \int_0^{y^+} \upsilon^+\, dy^+ . \tag{41.6}$$

It is assumed that the universal velocity profile prevails and that therefore υ^+ depends only on y^+ and hence ψ depends only on y^+, or only on υ^+, but not on x. Therefore, the coordinate transformation takes the form

$$\left(\frac{\partial\Theta}{\partial \psi}\right)_x = \left(\frac{\partial\Theta}{\partial \upsilon^+}\right)_{x^+} \left(\frac{\partial \upsilon^+}{\partial \psi}\right)_x + \left(\frac{\partial\Theta}{\partial x^+}\right)_{\upsilon^+} \left(\frac{\partial x^+}{\partial \psi}\right)_x = \left(\frac{\partial\Theta}{\partial \upsilon^+}\right)_{x^+} \frac{d\upsilon^+}{dy^+} \bigg/ \frac{d\psi}{dy^+}$$

$$= \frac{1}{\nu\upsilon^+}\frac{d\upsilon^+}{dy^+}\left(\frac{\partial\Theta}{\partial \upsilon^+}\right)_{x^+} .$$

$$\left(\frac{\partial\Theta}{\partial x}\right)_\psi = \left(\frac{\partial\Theta}{\partial x^+}\right)_{\upsilon^+} \left(\frac{\partial x^+}{\partial x}\right)_\psi + \left(\frac{\partial\Theta}{\partial \upsilon^+}\right)_{x^+} \left(\frac{\partial \upsilon^+}{\partial x}\right)_\psi = \frac{\upsilon_*}{\nu}\left(\frac{\partial\Theta}{\partial x^+}\right)_{\upsilon^+} .$$

The diffusion equation becomes

$$\upsilon^+ \frac{\partial\Theta}{\partial x^+} = \frac{d\upsilon^+}{dy^+}\frac{\partial}{\partial \upsilon^+}\left[\frac{D + D^{(t)}}{\nu}\frac{d\upsilon^+}{dy^+}\frac{\partial\Theta}{\partial \upsilon^+}\right]. \tag{41.7}$$

If τ is taken to be constant across the diffusion-layer region where the universal velocity profile applies, then Eq. 35.6 can be used to eliminate the derivative dv^+/dy^+, and the diffusion equation finally becomes

$$v^+\left(1+\frac{\nu^{(t)}}{\nu}\right)\frac{\partial \Theta}{\partial x^+}=\frac{\partial}{\partial v^+}\left[\frac{D+D^{(t)}}{\nu+\nu^{(t)}}\frac{\partial \Theta}{\partial v^+}\right]. \tag{41.8}$$

The boundary conditions are

$$\Theta=0 \text{ at } v^+=0,\quad \Theta=1 \text{ at } v^+=\infty,\quad \Theta=1, \text{ at } x^+=0. \tag{41.9}$$

This final form of the diffusion equation has several advantages. It is expressed in terms of independent variables appropriate for the application of the universal velocity profile. Thus, since v^+ is a function of y^+ as given by Fig. 35.1, $1 + \nu^{(t)}/\nu$ as given by Fig. 35.2 can also be expressed as a function of v^+. In the same way, $\left(D+D^{(t)}\right)/\left(\nu+\nu^{(t)}\right)$ depends only on v^+ and the value of the Schmidt number. Figure 39.1 gives $D^{(t)}/\nu$ as a function of y^+. Equations 41.8 and 41.9 can, therefore, be solved for a given value of the Schmidt number, and this calculation is independent of the particular character of the hydrodynamic flow. Spalding has solved Eqs. 41.8 and 41.9 numerically for Sc = 1 [1]. See also Ref. [2].

Although Θ can be calculated as a function of v^+ and x^+ as outlined above, it is necessary to solve the hydrodynamic problem and, in particular, to determine the shear-stress variation along the wall in order to apply these results to a given problem. For example, the mass flux at the wall is given by

$$J_o=D\left.\frac{\partial \bar{c}_i}{\partial y}\right|_{y=0}=\frac{D}{\nu}\sqrt{\frac{\tau_o}{\rho}}(c_\infty-c_o)\left.\frac{\partial \Theta}{\partial v^+}\right|_{v^+=0}. \tag{41.10}$$

This approach to turbulent mass transfer is an improvement over the "analogy" treatments, which involve assumptions like

$$\bar{\tau}/\bar{J}_{iy}=\text{constant}=\tau_o/J_o \tag{41.11}$$

and thus remove the need for solving partial differential equations. The assumptions here are milder; it is assumed that the diffusion layer is so thin that both v^+ and $\nu^{(t)}/\nu$ are universal functions of y^+, and it is also assumed that $D^{(t)}=\nu^{(t)}$ or, at least, that $D^{(t)}/\nu$ is also a universal function of y^+.

Problems

41.1 The Spalding transformation allows us to treat mass transfer in turbulent flow in a general manner, which is independent of the details of the geometry of the system. We are told that the partial differential equation has been solved numerically for Schmidt numbers up to 1000, for the case where the concentration is a constant on the solid surface.

(a) Your assignment is to set the problem up so that we can solve the problem asymptotically in the limit of an infinite Schmidt number. Thus, you are to stretch variables in such a way that the Schmidt number disappears from the problem. In doing this, it is anticipated that we would be able to infer some results for asymptotic behavior without actually solving the problem numerically.

(b) What can you infer about an entry-length region in this problem?

(c) What can you infer about a far-downstream region in this problem?

(d) What can you say about physical factors such as the eddy diffusivity and the eddy viscosity, which might play a role in different regions, as examined in Parts (b) and (c)?

(e) Let the surface flux density be denoted J_o. Can you infer anything about the dependence of J_o on the diffusion coefficient D?

(f) Discuss whether the assumptions underlying the Spalding transformation are reinforced, violated, or are not affected by considerations of the limit of large Schmidt numbers.

41.2 The results of Van Shaw, Reiss, and Hanratty [3] indicate that the mass-transfer entry length in turbulent pipe flow ranges from 2 diameters to 0.5 diameter as the Reynolds number ranges from 5000 to 75,000. We assume that the situation referred to involves a fluid with large Schmidt number and a fully developed velocity profile (that is, the mass-transfer section of the pipe is substantially downstream from the entrance to the pipe).

(a) Apply the semi-empirical theory of the universal velocity profile to this problem by formulating a differential equation and boundary conditions for the concentration distribution in this region.

(b) Introduce coordinates appropriate to the theory of the universal velocity profile.

(c) Introduce approximations appropriate for large Schmidt number.

(d) Infer an order of magnitude for the entrance length for mass transfer by casting the problem in a form independent of the Reynolds number and the Schmidt number.

References

1. D. B. Spalding. "Heat Transfer to a Turbulent Stream from a Surface with a Step-wise Discontinuity in Wall Temperature." *International Developments in Heat Transfer. Part II*, pp. 439–446. New York: The American Society of Mechanical Engineers, 1961.
2. J. Kestin and P. D. Richardson. "Heat Transfer across Turbulent, Incompressible Boundary Layers." *International Journal of Heat and Mass Transfer*, **6**, 147–189 (1963).
3. P. Van Shaw, L. Philip Reiss, and Thomas J. Hanratty. "Rates of Turbulent Transfer to a Pipe Wall in the Mass Transfer Entry Region." *AIChE Journal*, **9**, 362–364 (1963).

Chapter 42

New Perspective in Turbulence

Turbulence is a statistical phenomenon. It should be possible to develop a theory which is expressed by *local* relationships among several statistical quantities. Candidate statistical quantities include eddy viscosity, total (local) stress, and volumetric dissipation $\mathcal{D}_V = -\tau{:}\nabla\mathbf{v}$. This is referred to as the dissipation theorem.

In pipe flow, the theorem can be in harmony with the 1932 experiments [1, 2] in showing that the eddy kinematic viscosity rises from zero at the wall, passes through a maximum, and declines almost but not quite to zero on the axis. This contrasts with Figs. 35.1 and 36.1, which show peaked velocity and a total kinematic viscosity going to $-\nu$ on the axis. It also contrasts with Fig. 35.2, which shows a maximum value of the total viscosity on the axis.

The dissipation theorem has been applied to rotating cylinders [3, 4] and to pipe flow [2, 5]. It is also being applied to two systems where the flow develops as the fluid flows by the surface, that is, to the flat plate at zero incidence and to the rotating disk. In these two systems, the flow passes through laminar, transition, and turbulent flow. In the first two systems, the stress distribution is known in advance from a simple momentum balance. In the latter two systems, the stress is not known in advance, and one needs to solve the momentum equation simultaneously with the dissipation theorem.

The Newman Lectures on Transport Phenomena
John Newman and Vincent Battaglia

ISBN 978-981-4774-27-7 (Hardcover), 978-1-315-10829-2 (eBook)
www.jennystanford.com

Murphree [6] and Levich [7, 8] postulated a viscous sublayer very near the surface where the eddy viscosity declines with the cube of the distance from the surface and the eddy diffusivity is proportional to y^3 or y^4. See Chapters 39 and 40. One can perhaps resolve this controversy [9] by noting that Fig. 39.1 shows evidence of both slopes. However, at high Sc, the diffusion layer lies well within the viscous sublayer, and the evidence of a y^4 behavior as expressed in the table below Eq. 39.3 is very difficult to see experimentally.

Nevertheless, Vorotyntsev et al. [10] show that the eddy diffusivity is substantially different from the eddy kinematic viscosity in the viscous sublayer, even though they may be very similar in the external turbulent flow. One can assume that Fig. 39.1 applies to the eddy diffusivity at some unspecified but moderately high Sc, showing hints of both a y^3 dependence and a y^4 dependence. On the other hand, the profile of the eddy kinematic viscosity should be a straight line in the viscous sublayer with a slope of 3 for all values of y^+ less than about 20. Furthermore, the coefficient of y^3 in the y^3 region for $\mathcal{D}(y)$ can be related to the coefficient of y in the y-region outside the viscous sublayer [11].

References

1. J. Nikuradse. "Gesetzmässigkeiten der turbulentem Strömung in glatten Rohren." *Forschungsheft 356*, Beilage zu *Forschung auf dem Gebiete des Ingenieurwesens*, Edition B, Volume 3. Berlin NW7: VDI-Verlag GMBH, 1932. Translated as *J. Nikuradse, Laws of Turbulent Flow in Smooth Pipes*, NASA TT F-10, 359. Washington: National Aeronautics and Space Administration, October 1966.
2. John Newman. "Further Thoughts on Turbulent Flow in a Pipe." *Russian Journal of Electrochemistry*, **55**, 34–43 (2019).
3. John Newman. "Theoretical Analysis of Turbulent Mass Transfer with Rotating Cylinders." *Journal of the Electrochemical Society*, **163**, E191–E198 (2016).
4. John Newman., "Turbulent Flow with the Inner Cylinder Rotating." *Russian Journal of Electrochemistry*, **55**, 44–51 (2019).
5. John Newman. "Application of the Dissipation Theorem to Turbulent Flow and Mass Transfer in a Pipe." *Russian Journal of Electrochemistry*, **53**, 1061–1075 (2017).

6. E. V. Murphree. "Relation between Heat Transfer and Fluid Friction." *Industrial and Engineering Chemistry*, **24**, 726–736 (1932).
7. B. Levich. "The Theory of Concentration Polarization, I." *Acta Physicochimica U.R.S.S.*, **17**, 257–307 (1942).
8. B. Levich. "The Theory of Concentration Polarization, II." *Acta Physicochimica U.R.S.S.*, **19**, 117–132 (1944).
9. John Newman. "Eddy Diffusivity in the Viscous Sublayer." *Russian Journal of Elecrochemistry*, **55**, 1031–1033 (2019).
10. M. A. Vorotyntsev, S. A. Martem'yanov, and B. M. Grafov. "Closed Equation of Turbulent Heat and Mass Transfer." *Journal of Experimental and Theoretical Physics*, **52**, 909–914 (1980).
11. John Newman. "Viscous Sublayer." *Russian Journal of Electrochemistry*, **56**, 263–269 (2020).

Suggested Reading for Turbulent Flow

R. Byron Bird, Warren E. Stewart, and Edwin N. Lightfoot. *Transport Phenomena*, 2nd Edition. New York: John Wiley & Sons, 2002, pp. 152–172, 407–414, and 657–667.

Hermann Schlichting, *Boundary-Layer Theory*, 7th Edition, translated by J. Kestin. New York: McGraw-Hill Book Company, 1979, pp. 39–42, etc.

Veniamin G. Levich, *Physicochemical Hydrodynamics*. Englewood Cliffs, N. J.: Prentice-Hall, Inc., 1962, pp. 20–36, 139–183.

J. Kestin and P. D. Richardson, "Heat Transfer across Turbulent Incompressible Boundary Layers," *International Journal of Heat and Mass Transfer*, **6**, 147, 1963.

T. K. Sherwood, "Mass, Heat, and Momentum Transfer between Phases," *Chemical Engineering Progress Symposium Series*, **55**(25), 71, 1959.

Appendix A

Vectors and Tensors

You have seen that some of our equations, such as the equation of motion, have become quite lengthy. This frequently leads one to introduce vector notation, which has several advantages:

1. The equations become considerably more compact when written in vector notation.
2. The equations have significance independent of any particular coordinate system.
3. It is easier to grasp the meaning of an equation (after the vector notation becomes familiar).

For the present, you may regard vector notation as a form of shorthand writing. However, it would be a good idea for you to develop an intuitive feel for the significance of some of the more common vector operations.

A vector has both magnitude and direction and can be decomposed into components in three rectangular directions:

$$\mathbf{v} = \mathbf{e}_x v_x + \mathbf{e}_y v_y + \mathbf{e}_z v_z.$$

The *divergence* of a vector field is

$$\nabla \cdot \mathbf{v} = \frac{\partial v_x}{\partial x} + \frac{\partial v_y}{\partial y} + \frac{\partial v_z}{\partial z}.$$

This quantity is a scalar whose physical significance can be seen most easily from the continuity equation

$$\frac{\partial \rho}{\partial t} = -\nabla \cdot (\rho \mathbf{v}).$$

The mass flux is $\rho\mathbf{v}$, showing the direction and magnitude of mass transfer per unit area, and $\nabla \cdot (\rho\mathbf{v})$ represents the "rate of mass flowing away from a point." Hence the name "divergence." We might call $-\nabla\cdot(\rho\mathbf{v})$ the "convergence" of the mass flux $\rho\mathbf{v}$. Then the equation of continuity says

$$\frac{\partial \rho}{\partial t} = -\nabla\cdot(\rho\mathbf{v})$$

Rate of accumulation = Rate of convergence or net input.

The *curl* of a vector field yields another vector.

$$\boldsymbol{\Omega} = \nabla\times\mathbf{v} = \begin{vmatrix} \mathbf{e}_x & \mathbf{e}_y & \mathbf{e}_z \\ \dfrac{\partial}{\partial x} & \dfrac{\partial}{\partial y} & \dfrac{\partial}{\partial x} \\ v_x & v_y & v_z \end{vmatrix}$$

$$= \mathbf{e}_x\left(\frac{\partial v_z}{\partial y} - \frac{\partial v_y}{\partial z}\right) + \mathbf{e}_y\left(\frac{\partial v_x}{\partial z} - \frac{\partial v_z}{\partial x}\right) + \mathbf{e}_z\left(\frac{\partial v_y}{\partial x} - \frac{\partial v_x}{\partial y}\right).$$

When $\mathbf{v}$ is the fluid velocity, $\boldsymbol{\Omega}$ is known as the "vorticity," which may be regarded as the angular velocity (radians/second) of a fluid element. This vector operation will be used very seldom (if at all) in this course.

The *gradient* of a scalar field is a vector.

$$\nabla\rho = \mathbf{e}_x\frac{\partial \rho}{\partial x} + \mathbf{e}_y\frac{\partial \rho}{\partial y} + \mathbf{e}_z\frac{\partial \rho}{\partial z}.$$

The gradient of ρ shows the change of density with position. The direction of the gradient shows the direction of the greatest change, and the magnitude is the rate of change in this direction.

The gradient of a vector field, on the other hand, is a *tensor*. It has nine components because it is necessary to describe the rate of change of each component of velocity in each of three directions.

$$\nabla\mathbf{v} = \begin{pmatrix} \dfrac{\partial v_x}{\partial x} & \dfrac{\partial v_y}{\partial x} & \dfrac{\partial v_z}{\partial x} \\ \dfrac{\partial v_x}{\partial y} & \dfrac{\partial v_y}{\partial y} & \dfrac{\partial v_z}{\partial y} \\ \dfrac{\partial v_x}{\partial z} & \dfrac{\partial v_y}{\partial z} & \dfrac{\partial v_z}{\partial z} \end{pmatrix}.$$

A tensor is an operator for vectors. The result of a tensor operating on a vector is another vector:

$$\begin{aligned}\mathbf{T}\cdot\mathbf{a} &= \mathbf{e}_x(T_{xx}a_x + T_{xy}a_y + T_{xz}a_z)\\ &+\mathbf{e}_y(T_{yx}a_x + T_{yy}a_y + T_{yz}a_z)\\ &+\mathbf{e}_z(T_{zx}a_x + T_{zy}a_y + T_{zz}a_z).\end{aligned}$$

The stress $\boldsymbol{\tau}$ due to viscous forces is a tensor. Its nine components tell us the force acting on surfaces with various orientations.

$$\mathbf{f} = \mathbf{n}\cdot\boldsymbol{\tau}$$

where **n** is a unit vector normal to a surface and **f** is the stress on the surface.

A few definitions and identities are appended. The readers may find these useful from time to time in their domestic affairs.

Vector and tensor algebra and calculus

1. Definitions
   (a) Dyadic product $(\mathbf{a}\,\mathbf{c})_{ij} = a_i c_j$. (**a c** is a tensor.)
   (b) Double dot product $\boldsymbol{\sigma}:\boldsymbol{\tau} = \sum_i\sum_j \sigma_{ij}\tau_{ji}$.
   (c) A tensor operating on a vector from the right yields a vector.
   $$\mathbf{a}\cdot\boldsymbol{\tau} = \sum_i\sum_j \mathbf{e}_i a_j \tau_{ji}.$$
   (d) Transpose of a tensor. $(\boldsymbol{\tau}^*)_{ij} = \tau_{ji}$ or $\boldsymbol{\tau}\cdot\mathbf{a} = \mathbf{a}\cdot\boldsymbol{\tau}^*$.
   (e) Product of two tensors.
   $$(\boldsymbol{\tau}\cdot\boldsymbol{\sigma})\cdot\mathbf{v} = \boldsymbol{\tau}\cdot(\boldsymbol{\sigma}\cdot\mathbf{v}) \text{ or } (\boldsymbol{\tau}\cdot\boldsymbol{\sigma})_{ij} = \sum_k \tau_{ik}\sigma_{kj}.$$
   (f) The divergence of a tensor is a vector.
   $$\nabla\cdot\boldsymbol{\tau} = \sum_i\sum_j \mathbf{e}_i \frac{\partial\tau_{ji}}{\partial x_j}.$$
   (g) Laplacian of a scalar. $\nabla^2\Phi = \nabla\cdot\nabla\Phi = \sum_i \frac{\partial^2\Phi}{\partial x_i^2}$.
   (h) Gradient of vector. $(\nabla\mathbf{v})_{ij} = \partial v_j/\partial x_i$.

(i) Laplacian of a vector.

$$\nabla^2\mathbf{v} = \nabla\cdot\nabla\mathbf{v} = \nabla(\nabla\cdot\mathbf{v}) - \nabla\times\nabla\times\mathbf{v}.$$

2. Algebra
   (a) $\boldsymbol{\tau}:(\mathbf{a}\,\mathbf{b}) = \mathbf{b}\cdot(\boldsymbol{\tau}\cdot\mathbf{a})$.
   (b) $(\mathbf{u}\mathbf{v}):(\mathbf{w}\mathbf{z}) = (\mathbf{u}\mathbf{w}):(\mathbf{v}\mathbf{z}) = (\mathbf{u}\cdot\mathbf{z})(\mathbf{v}\cdot\mathbf{w})$.
   (c) $\mathbf{a}\cdot(\mathbf{b}\mathbf{c}) = (\mathbf{a}\cdot\mathbf{b})\mathbf{c}$.
   (d) $(\mathbf{a}\mathbf{b})\cdot\mathbf{c} = \mathbf{a}(\mathbf{b}\cdot\mathbf{c})$.
   (e) $\mathbf{a}\times(\mathbf{b}\times\mathbf{c}) = \mathbf{b}(\mathbf{a}\cdot\mathbf{c}) - \mathbf{c}(\mathbf{a}\cdot\mathbf{b})$.
   (f) $\mathbf{u}\cdot(\mathbf{v}\times\mathbf{w}) = \mathbf{v}\cdot(\mathbf{w}\times\mathbf{u})$.
   (g) $(\mathbf{u}\times\mathbf{v})\cdot(\mathbf{w}\times\mathbf{z}) = (\mathbf{u}\cdot\mathbf{w})(\mathbf{v}\cdot\mathbf{z}) - (\mathbf{u}\cdot\mathbf{z})(\mathbf{v}\cdot\mathbf{w})$.
   (h) $\mathbf{v}\cdot(\boldsymbol{\tau}^*\cdot\mathbf{w}) = \mathbf{w}\cdot(\boldsymbol{\tau}\cdot\mathbf{v})$.
3. Differentiation of products
   (a) $\nabla\phi\psi = \phi\nabla\psi + \psi\nabla\phi$ (a vector).
   (b) $\nabla\phi\mathbf{v} = \phi\nabla\mathbf{v} + (\nabla\phi)\mathbf{v}$ (a tensor).
   (c) $\nabla(\mathbf{a}\cdot\mathbf{c}) = \mathbf{a}\cdot\nabla\mathbf{c} + \mathbf{c}\cdot\nabla\mathbf{a} + \mathbf{a}\times\nabla\times\mathbf{c} + \mathbf{c}\times\nabla\times\mathbf{a}$
   $= (\nabla\mathbf{c})\cdot\mathbf{a} + (\nabla\mathbf{a})\cdot\mathbf{c}$ (a vector).
   (d) $\nabla\cdot(\phi\mathbf{v}) = \phi\nabla\cdot\mathbf{v} + \mathbf{v}\cdot\nabla\phi$ (a scalar).
   (e) $\nabla\cdot(\mathbf{v}\times\mathbf{w}) = \mathbf{w}\cdot(\nabla\times\mathbf{v}) - \mathbf{v}\cdot(\nabla\times\mathbf{w})$ (a scalar).
   (f) $\nabla\times(\phi\mathbf{v}) = \phi\nabla\times\mathbf{v} + (\nabla\phi)\times\mathbf{v}$ (a vector).
   (g) $\nabla\times(\mathbf{b}\times\mathbf{c}) = \mathbf{b}(\nabla\cdot\mathbf{c}) - \mathbf{c}(\nabla\cdot\mathbf{b}) + \mathbf{c}\cdot\nabla\mathbf{b} - \mathbf{b}\cdot\nabla\mathbf{c}$ (a vector).
   (h) $\nabla\cdot(\mathbf{a}\mathbf{b}) = (\nabla\cdot\mathbf{a})\,\mathbf{b} + \mathbf{a}\cdot\nabla\mathbf{b}$ (a vector).
   (i) $\nabla\cdot(\phi\boldsymbol{\tau}) = \phi\nabla\cdot\boldsymbol{\tau} + (\nabla\phi)\cdot\boldsymbol{\tau}$ (a vector).
   (j) $\nabla\cdot(\mathbf{u}\cdot\boldsymbol{\tau}) = \boldsymbol{\tau}:\nabla\mathbf{u} + \mathbf{u}\cdot\nabla\cdot\boldsymbol{\tau}^*$ (a scalar).
4. Various forms of Gauss' law (divergence theorem) and Stokes law. (dS = area element, $d\ell$ = line element, dv = volume element)
   (a) $\oint \mathbf{dS}\cdot\mathbf{F} = \int dv\nabla\cdot\mathbf{F}$.
   (b) $\oint \mathbf{dS}\phi = \int dv\nabla\phi$.
   (c) $\oint (\mathbf{dS}\cdot\mathbf{G})\mathbf{F} = \int dv\mathbf{F}\nabla\cdot\mathbf{G} + \int dv\mathbf{G}\cdot\nabla\mathbf{F}$.
   (d) $\oint \mathbf{dS}\times\mathbf{F} = \int dv\nabla\times\mathbf{F}$.

(e) $\oint d\mathbf{S}\cdot\boldsymbol{\tau}=\int dv\nabla\cdot\boldsymbol{\tau}.$

(f) $\oint d\mathbf{S}\cdot(\psi\nabla\phi-\phi\nabla\psi)=\int dv(\psi\nabla^2\phi-\phi\nabla^2\psi).$

(g) $\oint d\mathbf{l}\cdot\underline{F}=\int d\mathbf{S}\cdot\nabla\times\mathbf{F}.$

(h) $\oint d\mathbf{l}\phi=\int d\mathbf{S}\times\nabla\phi.$

5. Miscellaneous
   (a) $\nabla\cdot\nabla\times\mathbf{E}=0.$
   (b) $\nabla\times\nabla\phi=0.$
   (c) $\mathbf{w}\cdot\nabla\mathbf{v}=\sum_i\sum_j\mathbf{e}_i w_j\,\partial v_i/\partial x_j.$
   (d) $D/Dt=\partial/\partial t+\mathbf{v}\cdot\nabla.$
   (e) $\dfrac{D\mathbf{v}}{Dt}=\dfrac{\partial\mathbf{v}}{\partial t}+\nabla\left(\dfrac{1}{2}v^2\right)-\mathbf{v}\times\nabla\times\mathbf{v}.$

   (d) and (e): where $\mathbf{v}$ is the mass-average velocity.

Problems

A.1. Show that the Laplacian of a vector may be expressed as
$\nabla^2\mathbf{v}=\nabla\cdot\nabla\mathbf{v}=\nabla(\nabla\cdot\mathbf{v})-\nabla\times\nabla\times\mathbf{v}.$

A.2. Show that
$\nabla\cdot(\mathbf{a}\,\mathbf{b})=(\nabla\cdot\mathbf{a})\mathbf{b}+\mathbf{a}\cdot\nabla\mathbf{b}.$

Appendix B

Similarity Transformations

Similarity transformations are really quite useful because they allow partial differential equations, even nonlinear ones, to be reduced to ordinary differential equations. However, all similarity transformations cannot be discovered by the methods of dimensional analysis, and it is worthwhile to consider how to find a similarity transformation. (This appendix is taken from Chapter 21 of Ref. [1].)

Example of transient heat transfer in a slab

Let us look again at the problem

$$\frac{\partial T}{\partial t} = \alpha \frac{\partial^2 T}{\partial x^2}, \tag{B.1}$$

$$T = T_0 \text{ at } t = 0, \quad T = 0 \text{ at } x = 0, \quad T \to T_0 \text{ as } x \to \infty. \tag{B.2}$$

We seek a similarity transformation by assuming tentatively that

$$T = T(\eta) \quad \text{where} \quad \eta = x/h(t). \tag{B.3}$$

The similarity variable η introduced here can be seen to be fairly general. However, a more general form for the function would be

$$T = f(x) g(\eta), \tag{B.4}$$

because, as we saw in Problem 15.1 in Ref. [1], our transformation may look like

$$T = T_0 (1 - x/L) g(\eta). \tag{B.5}$$

But let us return to the form in equation 3. Substitution into equation 1 gives

$$\frac{\partial T}{\partial \eta}\left(-\frac{x}{h^2}\frac{dh}{dt}\right) = \alpha \frac{\partial^2 T}{d\eta^2}\frac{1}{h^2} \tag{B.6}$$

or

$$\eta \frac{dT}{d\eta}\left(\frac{h}{\alpha}\frac{dh}{dt}\right) + \frac{d^2T}{d\eta^2} = 0. \tag{B.7}$$

If the similarity transformation is to work, the variables t and x cannot appear separately in Eq. B.7 but only in the combination $\eta = x/h(t)$. This can be accomplished in the present case by setting the term in parentheses equal to a constant, where we choose to use 2 for simplification in later expressions. Thus, we obtain two ordinary differential equations:

$$\frac{d^2T}{d\eta^2} + 2\eta\frac{dT}{d\eta} = 0, \tag{B.8}$$

$$\frac{h}{\alpha}\frac{dh}{dt} = \frac{1}{2\alpha}\frac{dh^2}{dt} = 2, \tag{B.9}$$

with the solutions

$$T = A\int_0^{\eta} e^{-x^2}\,dx + B, \tag{B.10}$$

$$h^2 = 4\alpha t + C. \tag{B.11}$$

We should set $C = 0$ because we must also collapse two boundary conditions into one:

$$T \to T_0 \quad \text{as} \quad \eta \to \infty. \tag{B.12}$$

Then we have

$$\eta = x/2\sqrt{\alpha t}, \tag{B.13}$$

and the other boundary condition for Eq. B.8 or B.10 is

$$T = 0 \quad \text{at} \quad \eta = 0. \tag{B.14}$$

To satisfy the boundary conditions, Eq. B.10 becomes

$$T = T_0\frac{2}{\sqrt{\pi}}\int_0^{\eta} e^{-x^2}dx = T_0\,\text{erf}\left(x/2\sqrt{\alpha t}\right). \tag{B.15}$$

Example of transient diffusion to a growing sphere

Now let us consider mass transfer to a spherical drop whose radius increases with time in an arbitrary manner. The governing equation is the equation of convective diffusion 13.6, which reads in spherical coordinates

$$\frac{\partial c}{\partial t}+\upsilon_r\frac{\partial c}{\partial r}=\frac{D}{r^2}\frac{\partial}{\partial r}\left(r^2\frac{\partial c}{\partial r}\right)=D\frac{\partial^2 c}{\partial r^2}+\frac{2D}{r}\frac{\partial c}{\partial r}, \tag{B.16}$$

for spherical symmetry. For boundary conditions we take

$$c=0 \quad \text{at} \quad r=r_0(t) \quad \text{and} \quad c\rightarrow c_\infty \quad \text{as} \quad r\rightarrow\infty, \tag{B.17}$$

where r_0 is the radius of the drop and depends on time. We shall, at the outset, assume that the diffusion layer, in which the concentration deviates appreciably from the bulk value c_∞, is thin compared to the radius of the drop and that consequently the second term on the right in Eq. B.16 can be neglected.

The radial velocity is determined by the rate of growth of the drop:

$$4\pi r^2\upsilon_r=4\pi r_0^2 dr_0/dt \tag{B.18}$$

or

$$\upsilon_r=\frac{r_0^2}{r^2}\frac{dr_0}{dt}. \tag{B.19}$$

This is equivalent to the use of the continuity equation 13.2 for an incompressible fluid.

Now let us change to the variable

$$y=r-r_0(t), \tag{B.20}$$

the normal distance from the surface of the drop. The coordinate transformation from r, t to y, t gives

$$\left(\frac{\partial c}{\partial t}\right)_r=\left(\frac{\partial c}{\partial t}\right)_y\left(\frac{\partial t}{\partial t}\right)_r+\left(\frac{\partial c}{\partial y}\right)_t\left(\frac{\partial y}{\partial t}\right)_r=\frac{\partial c}{\partial t}-\frac{dr_0}{\partial t}\frac{\partial c}{\partial y}. \tag{B.21}$$

$$\left(\frac{\partial c}{\partial r}\right)_t=\left(\frac{\partial c}{\partial t}\right)_y\left(\frac{\partial t}{\partial r}\right)_t+\left(\frac{\partial c}{\partial y}\right)_t\left(\frac{\partial y}{\partial r}\right)_t=\frac{\partial c}{\partial y}. \tag{B.22}$$

$$\left(\frac{\partial^2 c}{\partial r^2}\right)_t=\frac{\partial^2 c}{\partial y^2}. \tag{B.23}$$

Hence, Eq. B.16 becomes

$$\frac{\partial c}{\partial t}-\frac{dr_0}{dt}\frac{\partial c}{\partial y}+\frac{r_0^2}{r^2}\frac{dr_0}{dt}\frac{\partial c}{\partial y}=D\frac{\partial^2 c}{\partial y^2}, \tag{B.24}$$

or

$$\frac{\partial c}{\partial t}-\left(1-\frac{r_0^2}{r^2}\right)\frac{dr_0}{dt}\frac{\partial c}{\partial y}=D\frac{\partial^2 c}{\partial y^2}, \tag{B.25}$$

or

$$\frac{\partial c}{\partial t}-\frac{(r-r_0)(r+r_0)}{r^2}\frac{dr_0}{dt}\frac{\partial c}{\partial y}=D\frac{\partial^2 c}{\partial y^2}. \tag{B.26}$$

Again on the basis of the thinness of the diffusion layer compared to the radius of the drop, we approximate $r + r_0$ by $2r_0$ and approximate r^2 by r_0^2, and Eq. B.26 becomes

$$\frac{\partial c}{\partial t}-\frac{2y}{r_0}\frac{dr_0}{dt}\frac{\partial c}{\partial y}=D\frac{\partial^2 c}{\partial y^2}. \tag{B.27}$$

Now seek a similarity solution of the form

$$c = c(\eta) \quad \text{where} \quad \eta = y/h(t). \tag{B.28}$$

Substitution into Eq. B.27 gives

$$\frac{\partial c}{\partial \eta}\left(-\frac{y}{h^2}\frac{dh}{dt}\right)-\frac{2y}{r_0}\frac{dr_0}{dt}\frac{\partial c}{\partial \eta}\frac{1}{h}=D\frac{\partial^2 c}{\partial \eta^2}\frac{1}{h^2} \tag{B.29}$$

or

$$\frac{d^2 c}{d\eta^2}+\frac{\eta}{D}\frac{dc}{d\eta}\left(h\frac{dh}{dt}+2\frac{h^2}{r_0}\frac{dr_0}{dt}\right)=0. \tag{B.30}$$

As in the case of Eq. B.7, the similarity transformation is successful only if the variables t and y appear in Eq. B.30 only in the combination $\eta = y/h(t)$. Consequently, we set the term in parentheses equal to a constant, 2D, and thereby generate two ordinary differential equations

$$\frac{d^2 c}{d\eta^2}+2\eta\frac{dc}{d\eta}=0, \tag{B.31}$$

$$h\frac{dh}{dt}+2\frac{h^2}{r_0}\frac{dr_0}{dt}=2D. \tag{B.32}$$

Equation B.32 is a linear, first order differential equation for h^2,

$$\frac{dh^2}{dt}+4\frac{h^2}{r_0}\frac{dr_0}{dt}=4D, \tag{B.33}$$

with the solution (see Chapter 2, Ref. [1])

$$h^2 r_0^4 = \int_0^t 4 D r_0^4 dt + C. \tag{B.34}$$

If we impose the initial condition

$$c = c_\infty \quad \text{at} \quad t = 0, \tag{B.35}$$

then we should set the integration constant C equal to zero so that this initial condition and the boundary condition at $r = \infty$ collapse into a single boundary condition

$$c \to c_\infty \quad \text{as} \quad \eta \to \infty, \tag{B.36}$$

and the similarity variable becomes

$$\eta = y r_0^2 \bigg/ \left[4D \int_0^t r_0^4 dt \right]^{1/2}. \tag{B.37}$$

The boundary condition at $r = r_0$ becomes

$$c = 0 \quad \text{at} \quad \eta = 0, \tag{B.38}$$

and the solution to Eq. B.31 becomes

$$c = c_\infty \frac{2}{\sqrt{\pi}} \int_0^\eta e^{-x^2} dx = c_\infty \operatorname{erf}(\eta), \tag{B.39}$$

where η is given by Eq. B.37.

This completes our solution of the problem of mass transfer to a spherical drop whose radius increases with time in an arbitrary manner. The similarity transformation B.37 could not be arrived at by dimensional arguments alone, and it is helpful to have an orderly procedure for searching for such transformations. In the course of obtaining the solution we made several approximations; the term $(2D/r)\partial c/\partial r$ was neglected in Eq. B.16, and y was neglected compared to r_0 in going from Eq. B.26 to Eq. B.27. It should be possible to justify these approximations *a posteriori* and to use the solution obtained above as a basis for obtaining higher order approximations. This is carried out in Ref. [2] for the case where the volume of the drop increases linearly with time.

Sedov [3] gives a variety of similarity solutions to partial differential equations.

Problems

B.1. Show that the total rate of mess transfer to a drop growing at a constant volumetric rate is

$$4\sqrt{\frac{7\pi D}{3}}\left(\frac{3Q}{4\pi}\right)^{2/3} c_\infty t^{1/6},$$

where Q is the volumetric rate of growth (cm^3/s). This is the result of Ilkovič for the rate of mass transfer to a drop growing at the tip of a capillary.

B.2. A gas bubble grows spherically in a supersaturated solution. Suppose that the convection is due entirely to the growth of the bubble, and show that the size of the bubble is given by

$$r_0 = 2\sqrt{\frac{3Dt}{\pi}}\frac{RT}{p}(c_\infty - c_0),$$

where t is the time since nucleation of the bubble, D is the diffusion coefficient of the gas dissolved in the solution, R is the gas constant, T is the absolute temperature, p is the pressure inside the bubble taken to be constant, c_∞ is the concentration of the gas in the supersaturated bulk solution, and c_0 is the saturation concentration prevailing at the surface of the bubble.

References

1. John Newman and Vincent Battaglia. *The Newman Lectures on Mathematics*. Singapore: Jenny Stanford Publishing (formerly Pan Stanford Publishing), 2018.
2. John Newman. "The Koutecký correction to the Ilkovič equation." *Journal of Electroanalytical Chemistry and Interfacial Electrochemistry*, **15**, 309–312 (1967).
3. L. I. Sedov. *Similarity and Dimensional Methods in Mechanics*. New York: Academic Press, 1959.

Index

acceleration 31, 219
aggregations of molecules 257, 258
angular velocity 101, 105, 145, 221, 268, 298
anisotropic crystal 54, 55
asymptotes 124, 129, 132, 133, 157, 160, 161, 169, 171, 203
average velocity 11, 12, 17, 19, 24, 27, 28, 43, 257, 265
axisymmetric 141, 143, 144, 161, 253
  flow 141, 143, 144, 161
  fluid 253

Bernoulli's equation 145
binary system 24, 25, 27, 33, 49, 184
Blasius series 147, 149, 169
Blasius solution 205, 233
Boltzmann constant 73
Boltzmann velocity 42
boundary layer 125, 136–138, 142, 144, 145, 153, 159, 154, 161, 165, 184, 185, 228, 239, 253, 269–273, 287–290
  approximation 140, 229
  axisymmetric 145
  simple 252
  thin 93
  two-dimensional 145, 155, 269, 287
  two-dimensional laminar 161
boundary-layer 93, 107, 138, 143–145, 147, 154–157, 160, 162, 173, 185, 187, 240, 270, 272, 273, 287
  equation 143–145, 154–156, 160, 162, 185, 187, 270, 272, 273, 287
  flow 107, 154, 155, 157, 173
  problem 93, 138, 144, 145, 147, 185, 240
bubble 91, 92, 98, 128, 162, 240, 307
bulk 23, 24, 183, 189, 307
  fluid 183
  motion 23, 24
  solution 189, 307
buoyancy effect 184

Carnot efficiency 66, 67
centrifugal effect 101, 142
Chilton–Colburn equation 279, 284, 285
Clausius–Dickel column 55
collision 28, 71
  binary 32
  molecular 75, 258
conduction 14, 40
  nonlinear 54
  steady 40
  thermal 91
continuity equation 14, 20, 96, 99, 102, 105, 137, 139, 142, 144, 163, 218, 219, 297, 298
convection 6, 14, 91, 124, 133, 183, 190, 208–210, 212, 307
  forced 183, 184
convective diffusion 84, 85, 123, 127–131, 153, 154, 156, 161, 162, 164, 165, 173, 174, 179, 207–210, 212, 214, 215, 251, 252
convective effect 154

convergence 68, 298
coordinates 59, 153, 165, 291
  boundary-layer 143, 146, 154, 185
  cylindrical 101, 106, 214, 218
  spherical 92, 96, 97, 99, 124, 128, 141, 305
correction factor 192–195
Couette flow 10, 88
curvature effect 142, 154

Debye length 56
diffusion 23, 24, 27–29, 32, 33, 44, 48–50, 53, 54, 59, 91, 92, 124, 126, 127, 129, 130, 133, 154–156, 159, 161–164, 176, 186, 187, 207–210, 212, 213, 229, 233, 240, 241, 288, 289, 305, 306
  approximation 127, 130
  axial 179, 180, 204
  coefficient 24, 25, 32–36, 38, 40, 42, 45, 49, 50, 129, 130, 133, 160, 162, 165, 224, 231
  connective 184
  equation 27, 29, 32, 33, 176, 212, 288, 289
  flux 24, 33, 57, 190
  law 24, 33, 49, 50
  layer 91, 92, 124, 126, 129, 130, 133, 155, 156, 159, 161–164, 176, 186, 187, 207, 210, 212, 213, 240, 241, 305, 306
  linear 189
  liquid-phase 92
  radial 208, 209
  tangential 208, 210
  thermal 25, 33, 35, 51, 75, 224
  transient 90, 91, 304
dilute-solution approximation 224
dissipative effect 63
disturbance 90, 252–254
divergence theorem 300
drag 31, 45, 98, 118, 119
driving force 24, 31–33, 37, 38, 41, 42, 51, 53, 54, 184, 223
Dufour effect 75
Dufour energy flux 40
Duhamel's superposition integral 199

eddy diffusivity 252, 275–278, 281, 282, 290, 294
eddy viscosity 252, 262–264, 266, 268, 271, 273, 285, 290, 293, 294

electrolyte theory 236
Ellis model 92
elongational flow 240
empirical expression 258, 262
energy equation 14, 15, 17, 18, 39, 47, 258
entropy flux 48
equilibrium 36, 53, 54, 70–73, 75
  dissociation 36
  gravitational 37
Euler 119, 135–138, 140, 142–144, 154, 160, 161
  constant 119
  equation 135–136, 142
  solution 136–138, 140, 143, 144, 160, 161
  velocity 154
evaporation 189, 217, 224–226
expansions 108, 109, 111, 112, 114, 115, 119, 127, 165, 179, 184
  composite 109, 113, 119
  inner 108, 109, 112, 113
  outer 108, 109, 111–114
  power-series 108, 155
  regular-perturbation 234, 236
  thermal 17

Faraday's constant 59
Fick's law 24, 25, 27, 33, 53

film model 223–225, 227, 231, 234, 238–240
  standard 239
  steady-state 224
fluctuations 58, 68, 72, 73, 75, 251, 255, 256, 259
  random 67, 251, 255
fluctuation theory 55
fluid motion 9, 105, 141, 143, 183, 184
flux density 199, 223, 225, 228–230, 236–238
Fourier series 67, 74
Fourier's law 4, 15, 48, 53
free convection 184, 185, 187, 239

Gauss' law 300
Gibbs–Duhem equation 32, 36, 47
Graetz series 176, 177, 179
Grashof number 184, 185

Hadamard–Rybczyński flow 240
Hagen–Poiseuille law 93
heat 4, 13–18, 39, 40, 48, 50, 53–55, 57–59, 61, 65–67, 75, 84, 133, 156, 175, 183, 189, 234
  conduction 4, 15, 48, 53, 55
  engine 55, 58, 59, 65–67
  flux 15, 39, 40, 50, 54, 57, 61, 75
  transfer 4, 13–18, 40, 54, 84, 133, 156, 175, 183, 189, 234
higher-order coefficient 167
hydrodynamic flow 289

incompressible fluid 10, 83, 84, 99, 101, 117, 257, 305
inertial effect 118
inviscid fluid 145
Joule effect 63
Joule heating 40

kinematic viscosity 43, 160, 228, 275, 293, 294
kinetic theory 223

laminar flow 92, 93, 164, 252, 253, 255, 271, 283
Laplace's equation 239, 240
Lennard– Jones "6-12" potential 44
Lennard–Jones parameters 46
lubrication approximation 217

one-dimensional approximation 180

macroscopic theory 4
macroscopic transport 74
microscopic theory 42
molar flux 23

Navier–Stokes equation 4, 95, 99, 128, 145
Nusselt number 21, 24, 33, 69, 124, 126, 127, 159, 162, 164, 177, 178, 188
  average 124, 126, 127, 159, 162, 164, 177, 178, 188
  molar 21, 24, 33, 69

Péclet numbers 124, 179, 210, 212
Peltier effect 63
Peltier heats 64
penetration model 223, 226, 227, 229, 230, 232, 235, 237, 239, 240
perturbation 119, 136, 235
  classical 107, 109
  singular 109, 115
perturbation expansion 107, 108, 115, 118, 236
polarography 240
pore 56, 180, 181
power-law fluid 92, 93

power series 104, 107, 108, 147, 167
Prandtl number 156
pressure 18, 20, 21, 25, 32, 42, 44, 46, 49, 51, 53, 59, 84, 91, 96, 98, 99, 105, 106, 120, 135, 138, 142, 144, 165. 184, 219, 224, 228, 254–256, 276
  diffusion 25, 224
  dimensionless 103
  distribution 98, 99, 105, 106, 120, 135, 165
  disturbance 254
  drag 276
  dynamic 84, 87, 105, 106, 121, 228
  gradient 51, 59, 84
  hydrostatic 84
  thermodynamic 84
  vapor 92
  variation 99, 138, 184, 219, 228

quasi-steady-state approximation 110

radial flux 283
reciprocal relation 54–58, 61, 74, 75, 224
Redlich–Kwong equation 115
refrigerator 55, 58, 64
relaxation time 27, 28, 31
Reynolds analogy 279
Reynolds number 95, 117–119, 124, 129, 153–155, 207, 213, 266–268, 271, 279, 282, 284, 285, 290, 291
rotating disk 101–106, 131–134, 145, 146, 163, 164, 192–195, 213, 214, 239, 293

Schmidt number 126, 129, 153–157, 160, 163, 164, 167, 184, 186, 187, 192–194, 212–214, 236, 237, 241, 276–279, 289–291
semi-empirical theory 273, 277, 281, 284, 290
shear stress 89, 93, 136, 140, 252, 255, 256, 261, 263, 265, 266, 270, 271, 273, 276
shear-thinning fluid 93
similarity transformation 89, 90, 92, 93, 125, 139, 158–160, 162, 164, 176, 186, 187, 226, 303–307
single component fluid 11
solute flux 51
stagnant fluid 88
stagnation point 142, 143, 147, 160, 161, 167, 239
Stanton number 276, 278, 279, 283–285
statistical theory 258
steady flow 87, 101, 135
Stefan–Maxwell equation 224
stoichiometric coefficient 238
Stokes 91, 95, 98, 107, 118, 119, 123, 127, 154, 164, 207, 213, 214, 241
  approximation 118
  drag 118
  flow 91, 95, 123, 127, 154, 164, 207, 213, 214, 241
  law 98, 119, 300
  solution 98, 107, 118, 119, 123
Stokes–Einstein relation 44
stress 9, 10, 54, 60, 89, 258, 263, 283, 293, 299
symmetry 44, 54, 56, 64, 70, 143, 147, 154, 162, 163
  axial 102, 128, 214
  flow 147
  spherical 305

Taylor expansion 236
Taylor series 147, 165
tensor 10, 297–301
thermal conductance 65
thermal conductivity 43, 45, 48, 50, 55, 61, 67, 84

thermal-diffusion coefficient 54, 55, 61
thermal diffusivity 70, 85
thermoelectric effect 55, 58, 59, 61, 63, 65
Thompson effect 63
torque 193–195
transport 4, 9, 24, 32, 33, 41, 42, 43, 45, 46, 53, 55, 56, 70, 75, 223
  coefficient 24, 32, 41, 42, 55, 56, 70, 75
  phenomena 41, 42, 45, 53, 55, 75
  property 4, 9, 32, 33, 42, 43, 45, 46, 55, 75, 223
turbulence 252, 253, 258, 293, 294
turbulent 106, 251–253, 255-258, 261, 262, 264–268, 275–278, 281, 287, 289, 290, 293, 294
  flow 106, 251–253, 255, 257, 258, 261, 262, 264–268, 275–278, 281, 287, 290, 293, 294
  fluctuations 256, 278
  mass transfer 278, 281, 283, 289
  mechanism 256, 257, 275
  momentum flux 255–257
  stress 258
two-dimensional flow 143, 144, 161

universal function 147–149, 167, 168, 263, 267, 268, 273, 289

velocity 8, 9, 11, 19, 20, 92, 93, 98, 99, 101, 104, 105, 120, 121, 135, 136, 143, 144, 163, 209, 210, 228, 229, 251–253, 255, 256, 270, 271
  angular 101, 105, 145, 221, 268, 298
  average 11, 12, 17, 19, 24, 27, 28, 43, 257, 265
  dimensionless 103, 104, 153, 228, 265
  interfacial 85, 189–191, 194, 195, 225, 227, 231, 237, 241
  mass-average 20, 27, 189, 190, 301
  non-vanishing 10
  radial 305
  tangential 136, 160, 162
  time-average 258
velocity distribution 92, 99, 121, 123, 128, 129, 145, 163, 252, 282
velocity gradient 89, 136
viscosity 9, 44–46, 48, 49, 105, 135, 136, 141, 144, 228, 230, 255, 258, 261, 262, 265–267
viscous 4, 7, 8, 14, 53, 54, 93, 217, 258, 271, 276, 299
  effect 141
  force 4, 7, 8, 14, 93, 217, 299
  stress 8, 53, 54, 258, 271, 276
von Kármán solution 146
von Kármán transformation 103, 105, 106, 146, 192
vorticity 96, 99, 119, 136, 298

zero incidence 93, 228, 293